**Bur**

'A brilliantly readable and absorbing analysis of the capitalist roots of climate breakdown, and an inspiring rallying cry for activists everywhere to work together to build a just, ecosocialist future.'
—Grace Blakeley, editor, *Futures of Socialism*

'Few people still deny that climate change is taking place, but who is to blame for the crisis? Chris Saltmarsh sets the record straight, explaining that the capitalist system is to blame, and the fight for climate justice offers a way out. This rousing book demonstrates that by joining in solidarity with others fighting for a new society, we can remake the world for everyone rather than just the wealthy few.'
—Ashley Dawson, Professor of Postcolonial Studies, Graduate Center, CUNY and author of *People's Power: Reclaiming the Energy Commons*

'A great contribution to unveiling the roots of our crisis, rich in storytelling drawing from Chris' deep experience in organising for a world that centres people and planet.'
—Harpreet Kaur Paul, Human rights lawyer

'From "generation climate" to a transformative Green New Deal, this is a sure guide through the politics of environmental breakdown and why radical ambition is our safest path forward.'
—Mathew Lawrence, co-author of *Planet on Fire: A Manifesto for the Age of Environmental Breakdown*; Director of the think tank Common Wealth

'Accurately identifies the scale of the crisis facing us and offers strategic ideas for how we respond – a rallying cry in book form.'
—Callum Cant, author of *Riding for Deliveroo*

'Pushes the British climate movement to go further in their demands for ecological justice. Unlike many books about climate breakdown, this book understands the political and economic system that is holding us to ransom, and has a good idea of how to change it.'
—Sam Knights, activist and editor of *This Is Not A Drill: An Extinction Rebellion Handbook*

'Deftly draws upon his experiences in the student and Labour Party climate movements to provide a compelling analysis of how the climate movement must urgently pivot to take the capitalist system head on or fail.'
—Gaya Sriskanthan, co-chair of Momentum

**Outspoken by Pluto**
Series Editor: Neda Tehrani

Platforming underrepresented voices; intervening in important political issues; revealing powerful histories and giving voice to our experiences; Outspoken by Pluto is a book series unlike any other. Unravelling debates on feminism and class, work and borders, unions and climate justice, this series has the answers to the questions you're asking. These are books that dissent.

Also available:

# Burnt

## Fighting for Climate Justice

Chris Saltmarsh

PLUTO PRESS

First published 2021 by Pluto Press
345 Archway Road, London N6 5AA

www.plutobooks.com

Copyright © Chris Saltmarsh 2021

The right of Chris Saltmarsh to be identified as the author of this
work has been asserted in accordance with the Copyright, Designs
and Patents Act 1988.

British Library Cataloguing in Publication Data
A catalogue record for this book is available from the British Library

ISBN   978 0 7453 4182 8      Paperback
ISBN   978 1 78680 849 3      PDF
ISBN   978 1 78680 850 9      EPUB

This book is printed on paper suitable for recycling and made from
fully managed and sustained forest sources. Logging, pulping and
manufacturing processes are expected to conform to the environmental standards of the country of origin.

Typeset by Stanford DTP Services, Northampton, England

Simultaneously printed in the United Kingdom and United States of
America

# Contents

# Acknowledgements

Many thanks to Neda Tehrani for making the initial contact that led to this book, for having the confidence in me to write it, as well as for giving me the freedom and support to produce something I can be proud of.

Thanks to Rosie Carter-Rich for the critical feedback and validating praise, and for your love and support. Thanks to my parents: Jane and Phil, and grandparents: Liz, Phil, Cynthia and Bernie, and Anna and Jacob for all your love and encouragement.

Thanks to everyone at People & Planet for giving me time off to write and for giving me so many of the formative experiences which have produced the ideas in this book.

Many thanks to the friends and comrades who took the time to read drafts and offer their praise, challenges and expert insights: Abbie Mulcairn, Josh, Chris Jarvis, Rosie Wright, Minesh Parekh, Jamie Woodcock, Jake Woodier, Toby McKenzie-Barnes and Brendan Montague.

Thanks to Rob Abrams, Scarlett Westbrook and Gabrielle Jeliazkov for your testimonies.

Thanks to the many other comrades (of which there are far too many to name) with whom I've shared a common struggle for justice over the years. This book is the product of so many hours of organising, actions, pub chats, frustrated rants, plotting and imagining.

Thanks to everyone who's had any involvement in Labour for a Green New Deal: my fellow co-founders, team leads and members, local group members, and supporters. You've provided

# ACKNOWLEDGEMENTS

the organisational vehicle for many of the ideas in this book. Together we've demonstrated their potency, popularity and total necessity.

The vast majority of this book was written during various COVID-19 lockdowns. Thank you to every key worker and volunteer who kept our society going at great personal risk, especially those who are no longer with us. The Black Lives Matter uprisings were the other major event during the writing of this book. My solidarity goes to all those struggling for racial justice, especially as the climate crisis continues to intensify the harms of structural racism and colonialism.

# Introduction: Climate crisis

This is a book about climate change, those who are truly to blame for it, radical solutions to it and our power to win them. When you think about climate change, you might feel scared, anxious or nothing at all. Stories about the latest UN Climate Summit might make the climate feel like an abstract issue you personally have no power over. When you see the latest hurricane ripping through a major city, you might switch off and try not to think about it. Or those extreme weather events might always be on your mind, inescapable as they keep you up at night. You might view climate change as something for hippies or just a middle-class concern. Or you might wish that everyone would wake up and take this patently existential crisis a lot more seriously: the planet is on fire for fuck's sake! You may or may not recycle. You may or may not eat meat. You may or may not take long-haul flights. You might feel compelled to block a road, lock yourself to an oil rig or get arrested to stop climate change. You might want to do something, but you don't know what or how.

Whatever you feel or do about climate change, this book is for you. In it you'll find a story of climate change that puts questions of justice front and centre. What is unjust about climate change? It comes down to a wealthy elite profiting from business-as-usual while ordinary people bear the costs of their recklessness. In this story of climate change, the ruling class are the villains. They uphold an economic system from which they benefit, while the world is literally on fire. We – ordinary people in every neighbourhood, town, city and country in the world – are the heroes. As the devastating impacts of climate change collide with decades of economic dispossession, we have a historic opportunity to transform our global political and economic systems to put people before profit. We can change our relationship to the world we live in and repair for the harms already inflicted by climate change. This book is a call to action to those ready to stand up for climate justice and an invitation to those who have never thought of themselves as activists. Because without you, we can't win.

Aged 16, I started environmental campaigning at college against all kinds of waste (food, water and paper), for meat-free Mondays and against unnecessary flying. In short, I was your classic insufferable moralising liberal environmentalist. I had nothing on the youth strikers who have taken to the streets, often at an even younger age, with monthly strikes for climate justice and demands as radical as a Green New Deal. By the age of 18, I became a climate justice activist. On my first day at the University of Sheffield, I joined the fossil fuel divestment campaign for the University to stop investing in fossil fuel companies. This was an education in organising around an explicitly anti-corporate climate politics. At the same time, I became involved in direct-action campaigns, including mass invasions of coal mines and blockading fracking sites by locking my arms in fortified tubes in the middle of the main road connecting Blackpool and Preston. At the same time, I was active in left-wing politics. I occupied university buildings in the free education movement opposing tuition fees and the marketi-sation of higher education. I joined campaigns for a living wage, against detention centres and deportations, for housing justice and solidarity with Palestine. Later, I joined the Labour Party as

Jeremy Corbyn's leadership opened the party up as a vehicle for radical politics.

Being involved in climate and leftist organising at the same time taught me limitations of both as well as how they can draw from each other, to win together. We need a climate politics which faces up to capitalism as the root cause of the crisis and offers a compelling alternative vision for society. We need a strong left which takes the challenges and opportunities of climate change seriously, by drawing the links between environmental issues and the those affecting peoples' everyday lives.

## Impacts

In September 2019, Matt Wrack, General Secretary of the Fire Brigades Union, gave a rousing speech in support of a socialist Green New Deal at the Labour Party Conference. He told delegates: 'Now for us, in my industry, this is an industrial matter today. It is a trade union matter today. Firefighters in the UK are dealing with the effects of climate change every day of their working lives in extreme weather events. It's an issue for firefighters across the globe.' Wrack goes on to give the 2013 example of Yarnell Hill fire in Arizona where 19 firefighters died. All were trade union members who died at work. Amid severe drought, a crew was tasked with fighting a 300-acre fire 850 feet up Weaver Mountains. A thunderstorm brewed, sucking in air, and then blowing it out with fire bellowing like a volcanic eruption. The rapidly changing weather ultimately cost those lives as the crew made their way down the mountain into thicker smoke before encountering a fire beneath them that had burned four miles in 20 minutes. They deployed their small aluminium tent fire shelters, wrestling them to the ground as the flames passed over them and the heat became unimaginable.

When the crew was finally found an hour and a half later, the paramedic first on the scene confirmed the fatalities.[1] Though a tiny proportion of deaths caused by climate change already, the deaths of those 19 firefighters illustrate that working-class people experience the worst of climate change, whether they are front line workers or in the communities they protect.

In the years since, captivating videos of wildfires in the US across social media have become an annual feature of summer. Residents displaced from their California homes film as they drive along highways flanked by almighty flames. The apocalyptic images adorning 24/7 news and Twitter feeds make unavoidable the reality that climate change is happening right now with devastating effects. An anonymous firefighter wrote in *This is Not a Drill*, Extinction Rebellion's handbook:[2]

> I'm not emotional about burning structures: you get used to that. What's painful is the way the fire affects people: when they're actively evacuating and scared and are told to leave everything behind because the fire is imminent. I see dead animals, horses, family dogs that were left behind.

> If you're a millionaire in Malibu, you can rebuild. But communities like Paradise [a town swallowed by fire in December 2018] are mostly older, retired, working class folks. They can't afford to bounce back.

When Greta Thunberg urges leaders to act with the refrain 'our house is on fire', this is literally true for many working-class

---

1  Kylie Dickman, 'The True Story of the Yarnell Hill Fire', *Outside*, https://outsideonline.com/1926426/19-true-story-yarnell-hill-fire#close (last accessed 20 January 2020).

2  Firefighter, California, 'There's fear now', *This is Not a Drill: The Extinction Rebellion handbook* (Milton Keynes: Penguin Random House UK, 2019), pp. 46–7.

people around the world. Their possessions, homes and communities are burned to a crisp because of climate change. Some people lose everything. Firefighters are on the front lines experiencing and witnessing first-hand the trauma of climate change, but it isn't just fires that are having that effect. A study by researcher Tamma Carleton has shown that crop-damaging changes in temperatures have resulted in increased suicide rates among India's farmers.[3] When temperatures rise, crops are less successful. This means that farmers don't earn as much money and become trapped in cycles of debt. In 2019, the BBC reported the story of Mallapa, a farmer in the state of Andhra Pradesh.[4] One day in August 2018, Mallapa left his home 'to buy groceries' but in fact walked into town to buy all the necessary things for his funeral before taking his life due to a debt of 285,000 rupees (£3,100) to banks. Mallapa had a peanut farm ruined by drought and his debt was worsened by falling crop prices. Mallapa is just one example of the 59,300 suicides by Indian farmers since 1980.[5] As temperatures rise and environments change faster than populations can adapt, lives are ruined. The striking thing about Mallapa's story is that his suicide was not born of a moment of desperation. The time he takes to arrange his own funeral proves that this was a considered choice when colonial-capitalism left no other option.

The psychological effects of debt are predominately individualised. The stress, anxiety and depression endured due to significant indebtedness often fall on you alone. They make you

---

3   Tamma A. Carleton, 'Crop-damaging temperatures increase suicide rates in India', *Proceedings of the National Academy of Sciences of the United States of America* 114, no. 33 (2017).

4   'The struggling farmer who planned his own funeral', *BBC News*, https://bbc.co.uk/news/world-asia-india-46644440 (last accessed 21 February 2020).

5   Carleton, 'Crop-damaging temperatures', PNAS.

feel isolated and powerless. They can be fatal. In the case of Hurricane Maria in 2017, the effects of climate change were felt collectively. Naomi Klein's book *The Battle for Paradise* describes how Puerto Ricans got organised in response to the devastation of the worst storm to ever hit the region. Klein writes of months without power and water, victims cut off from the rest of the island with roads blocked by fallen trees, people living by flashlight and dependent on FEMA (Federal Emergency Management Agency) food aid. Reading the devastation described in words alone is insufficient to understand what happened in Puerto Rico. Just as those images of flames engulfing Californian highways had such a viral impact, seeing photographs of the aftermath of Hurricane Maria takes you just a step closer to understanding the wrath that climate change is capable of inflicting. The most visceral images are those of roads not just blocked but ripped to pieces as if they were the collateral damage of a superhero movie fight scene: cars and buildings submerged by flooding, and hundreds of people sleeping in stadiums converted into emergency accommodation. In truth, this is the collateral damage of capitalism's merciless pursuit of profit at all costs.

## Miseducation

For many years, people in the UK have been miseducated about what climate change means. This is not to say that information about climate change is wholly inaccurate, but that it is incomplete. For example, a GCSE Geography textbook published by awarding body AQA tells a limited story of climate change.[6] The 'significant effects of climate change' presented on the first page of the chapter lead with glaciers, ice caps and Arctic sea ice

---

6    Simon Ross & Nick Rowles, *GCSE (9-1) Geography AQA* (Oxford: Oxford University Press, 2016), pp. 40–62.

melting. Where attention is given to human suffering, the reality is minimised. The textbook draws attention to small-island nations like the Maldives and Tuvalu being 'under threat from sea level rise' but does not elaborate on the existential nature of the threat to the people living there. In fact, the Maldives is predicted to be entirely submerged with its entire population made refugees by 2050. The textbook highlights the threat of sea levels rising by 1 metre by 2100 with agricultural land in Bangladesh, Vietnam, India and China most at risk. It is hard for the reader to grasp that this is not a distant threat coming at the end of the century. Extreme flooding is already devastating huge populations in these countries and beyond.

The textbook draws an equivalency between 'natural' and 'human' causes of climate change. Of course, it is true that the climate changes naturally. It is also true that climate change has seriously accelerated with the onset of burning fossil fuels and the greenhouse effect. The textbook claims: 'Many scientists believe that [the correlation between rising carbon emissions and rising global average temperatures] provides clear evidence that human activities are affecting global climates.'[7] A more accurate description would be that there is consensus around anthropogenic climate change across the scientific community. The BBC has been criticised for the platform it gives to climate deniers in the name of 'balance'. This has contributed to the systematic miseducation of the British public, with climate change presented as a question of scientific contention to be debated by equally legitimate 'experts', rather than a scientific reality contested only by the fringe and the corrupt. Fossil fuel companies like ExxonMobil have funded think tanks propagating climate denial, giving them cover to continue extracting

7   Ibid., p. 45.

7

the oil, gas and coal causing greenhouse gas emissions behind climate change. In September 2018, editorial staff at the BBC were finally told that 'you do not need a "denier" to balance the debate'.[8]

In the textbooks and on broadcast media, the impacts of climate change are abstracted as technical policy debates without a proportionate sense of the scale of the suffering and devastation already being endured. Both may offer a cursory acknowledgement of impacts including droughts, floods, tropical storms or lower crop yields, but it is impossible to understand what these impacts mean without their political context and without the stories of those with direct experience. To understand climate change, our stories must directly address the question of justice.

Using the framework of justice has brought environmental and climate impacts into the political realm. It recognises that while the ruling class have contributed the most to climate change, they generally experience its impacts the least. On the other hand, those who do experience climate impacts have generally done the least to contribute to it. This is the structural injustice of climate change, shaped by relations of class, colonialism and gender. The history of colonialism (the construction of Empire through the occupation and exploitation of land globally) has been one of imposing capitalism and fossil fuel extraction to profit capitalists. The creation and reproduction of this global system has left the Global South with the harshest climate impacts and insufficient resources to adapt. These harms are also gendered. 71% of people who die from climate impacts are

---

8   Damian Carrington, 'BBC admits "we get climate change coverage wrong too often", *Guardian*, https://theguardian.com/environment/2018/sep/07/bbc-we-get-climate-change-coverage-wrong-too-often (last accessed 21 February 2020).

women.[9] This is largely due to women's relative lack of access to the security of wealth and economic independence and relative likelihood to have caring responsibilities, reducing mobility. At the foundations of climate injustices is class. The inequitable impacts of climate change are distributed most reliably along the lines of wealth and economic power. The poorest and working-class – whether you live in California, Andhra Pradesh, Puerto Rico or anywhere else – have more in common in the face of climate change than executives of fossil fuel companies and other capitalists profiting from disaster.

## Climate crisis

The term climate crisis is used a lot, sometimes synonymously with climate change or climate injustice, but looking at the etymology of the word 'crisis' we see that it means *turning point* or *decision*. If climate change is the ecological process, and climate injustice brings those realities into the political realm, then the climate crisis is the moment we find ourselves in, and our opportunity to build something new.

The causes and the effects of climate change are sparsely distributed across space and time. We cannot draw direct causal links between burning specific lumps of coal in England in the 1890s and a particular hurricane in the Caribbean today. But we do know that all emissions over time cumulate, creating the conditions for a changing climate. Some impacts creep up on us, like the slow depletion of animal populations or gradual heating of summer months. There is a worry that if those of us

9  Patricia Scotland, 'Women shouldering the burden of climate crisis need action, not speeches', *Guardian*, https://theguardian.com/global-development/2020/mar/13/women-shouldering-the-burden-of-climate-crisis-need-action-not-speeches (last accessed 14 December 2020).

relatively insulated from the most extraordinary climate impacts will slowly get used to the changing climate, we might not take advantage of the crisis – before it's too late. However, a combination of factors in recent years are shaking us out of this inertia. With every passing year, extreme weather events become more frequent and severe with the havoc of wildfires and hurricanes adorning broadcast and social media. Extreme weather becomes disaster when these events collide with the capitalist pressures of austerity, privatisation and dispossession. The neoliberal erosion of state capacity and social fabric leaves us even more vulnerable in the face of sudden climatic changes.

The effects of this are increasingly unavoidable, especially as protest movements like Extinction Rebellion and youth strikes push them to the top of the news agenda. In October 2018, before this new wave of climate protest emerged, the UN Intergovernmental Panel on Climate Change (IPCC) released a report detailing measures needed to meet the Paris Agreement's targets to limit global average temperature rises to 1.5 or 2 degrees. The media headline was that we had just twelve years to stop climate change. This fuelled growing climate angst kicked off by the release of a report in August 2018, warning of the possibility of a 'Hothouse Earth' (about as colourful as scientists may be expected to get with their descriptive language) where our planetary tipping points are reached much quicker than expected. Slowly but surely, we are understanding these disasters not as isolated humanitarian events but as part of a trend of climate change. And we are feeling an urgency to do something.

The climate crisis presents us with an opportunity to move beyond capitalism towards a new way of organising the economy. If our economy were owned and democratically controlled by workers and the public, sectors like energy, water, transport, housing, food, health and social care, education and manufacturing could all be run for the benefit of society and not shareholder profits. Together we could manage the transition to a zero-carbon society quickly and fairly. We could fairly adapt our society and economy to extreme climate changes that are now inevitable. If we guaranteed a real living wage for all workers, free education from cradle-to-grave, food guaranteed as a human right, high-speed rail connecting cities across borders, well-resourced emergency services and collective insurance for disaster, we would build a dignified and prosperous society that was designed to respect ecological limits and human rights globally. If we remade the institutions of global politics and decoupled the accumulation of capital from economic prosperity, we could begin empowering peoples and nations subject to the domination of colonial powers for many years. With people able to act in their own interests, they might resist foreign states and corporations exploiting their natural resources, stealing their land, and destroying their environments.

This vision is not really a radical one. It is necessary to overcome the climate crisis. Building this world from deep within capitalism's death-drive will not be easy, but it is possible. Every political vision needs a strategy to make it so. This book exposes the centrality of capitalism to the climate crisis and the importance of justice in our response. It asks why the climate movement has had limited success so far and whether new grassroots movements can be a source of hope. I make the case for a socialist climate justice politics oriented around a Green New Deal and propose a strategy for achieving it into the 2020s and beyond.

# Chapter 1

# The c-word (capitalism)

The idea that humans are collectively responsible for our climate and wider ecological crises is a popular one. In the trailer for his 2020 Netflix blockbuster, *A Life on Our Planet*, David Attenborough proclaims: 'The living world is a unique and spectacular marvel. Yet, the way we humans live on Earth is sending it into a decline. Human beings have overrun the world.' In backlash to well-funded climate denial, a human-centred climate terminology has come to dominate. Climate change is 'man-made', emissions are 'anthropogenic', our new geological epoch is the 'Anthropocene'. But neither obstructive climate change denial nor the crude homogenisation of humanity as an equally responsible collective agent are helpful frameworks. The root cause of climate change is in our system of organising the economy and our relationship to nature: capitalism.

The concept of the Anthropocene gained prominence in the 2000s. Its story is simple. We are living in a new geological epoch defined by human impact on the planet. Agriculture, industry, fossil fuel extraction and deforestation: these are core to the development of human civilisation. They have led to mass extinction, pollution, biodiversity collapse and climate change. In *Capitalism in the Web of Life*, Jason W. Moore problematises this narrative, arguing that it excludes considerations of class, capitalism, imperialism and culture as potential driving forces behind those ecological crises. The Anthropocene narrative 'does not challenge the naturalized inequalities, alienation, and violence inscribed in modernity's strategic relations of power and production. It is an easy story to tell because it does not ask us to think about these relations at all.'[1] Moore is saying that our economy has inequality and injustice so deeply ingrained that we cannot claim that the impoverished, displaced, exploited and terrorised are as responsible for climate change as those who govern and profit from this system. Are the Indigenous peoples resisting new oil pipelines as culpable as the companies and governments using military violence to force them through? Are subsistence farmers as blameworthy as multinational agri-businesses forcing farmers into debt and spreading industrial agriculture? Moore proposes an alternative name for our geological epoch: Capitalocene. We shouldn't get hung up on terminology, but Moore's intervention highlights that the Anthropocene is not just a scientific claim. It's an ideological story implicating those with no part in making the crisis as much as capitalists who continue to profit from climate injustice, and the capitalist system they operate through.

## The enemy is profit

In the most general terms, capital is money used to generate more money. Capitalism gets its name from the centrality of the accumulation of capital to the system. A capitalist (an individual or corporation) buys someone else's time to work for them. The capitalist then buys the materials, machinery and computers, and puts these to work to produce some commodity (such as, pasta, video games, solar panels or apartment buildings). The commodity is then sold for more money than was originally

---

1   Jason W. Moore, *Capitalism in the Web of Life: Ecology and the Accumulation of Capital* (London & New York: Verso Books, 2015), p. 170.

invested – this is profit (surplus value, in Marxist terms). The surplus value comes from what Marxists call the exploitation of labour. The capitalist pays the worker less than the value of what they have produced, so the capitalist can steal some for themselves. The profit is then reinvested to create more profits, and the cycle of capital accumulation continues. Capitalism structures the economy so that this process governs all activity. Any other considerations – justice, climate stability or human rights – are subordinated to the capital's accumulation. That's not to say those other interests are entirely ignored, but according to capitalist logic, maximising profit is the primary motive.

Capitalist ideology says that this process of accumulation is how society progresses. Capital creates wealth, which 'trickles down'. It is shared by everyone as it's distributed to workers through wages, taxed by the government and reinvested by capitalists into socially useful things. This rosy story isn't quite the reality, though. The profit motive usually gets in the way. To maximise returns, capitalists suppress workers' wages and invest in making the most profitable commodities, rather than what is socially needed. Capital constantly innovates new methods of tax avoidance through offshore accounts, legal loopholes and philanthropy. There's no point at which capitalists are satisfied with their wealth. There's no point at which profit can be subordinated to decarbonisation, restoring ecosystems, investing in public infrastructure or adapting to climate change. When capital is left to its own devices, there will always be a trade-off between profits and climate justice, and profit will always win out. Sometimes capital is restrained by the state when it does things that hurt capital collectively. We see this with minimum labour standards and basic environmental regulations, for example around the disposal of hazardous waste. These are generally the result of hard-fought organising by workers and activists.

As well as the capital accumulation of exploiting workers' labour, Moore explains the importance of what he calls 'appropriation'. This is the *unpaid* extraction of value from: labour (for example, unpaid housework), food, energy and raw materials.[2] The appropriation of these natures is necessary to maximisising profits by keeping production costs down. Central to capital accumulation, then, is the process of commodifying nature: turning natures into things to be bought, sold or used for production. Rare earth minerals, oceans, rainforests, fertile land and untapped fossil fuels don't have some essential monetary value. Value is imposed as capital expands its sphere of accumulation and workers transform natures into commodities, then sold for profit. Capitalism facilitates capital's insatiable drive for profit, resulting in rainforests cleared to sell the wood and expand agricultural space; oceans trawled for fish populations; mines created to extract minerals; and fossil fuels extracted to be burned for energy. Mainstream economists call these side effects 'negative externalities', but they are much more than the unfortunate costs of an otherwise efficient system. These appropriations have driven us into existential ecological crises, including climate change.

Although some debates present a conflict of interests between workers and environment, workers and environment are in fact united by their centrality to the process of capital accumulation. The exploitation of human labour and appropriation of nature mutually allow capital to profit from its investments. With neither in the popular interest, the need for wage labour and the commodification of nature to be invented and then violently

---

2    Moore, *Web of Life*, p. 17.

imposed by capital is another case of unity. David Harvey describes how the law was used to standardise the wage relation, the length of the working day and to criminalise beggars and vagabonds.[3] Under early-capitalism, peasants were expelled from their land to force them into urban areas so that they could be exploited as workers, simultaneously making way for their land to be appropriated by capital. Methods of imposing this system on workers and nature have evolved over time, including through: 'enclosures, punitive rent hikes, land clearings, the introduction of agricultural machinery, crushing competition from agribusiness, military confiscations, bans on inheritance of small plots or some other blow that makes continued life on the land impossible.'[4] The profit motive has driven capital to intensify the exploitation of labour, forcing workers into worse conditions, and the unsustainable commodification of nature in the extraction of fossil fuels, hurtling the climate past irreversible tipping points. Workers and those vulnerable to climate injustices share an interest in transforming the economy so that human well-being, justice and a stable climate are prioritised over capital's profits.

## Capitalism's endurance

The by-products of an economy oriented around profit are plain for all to see: abhorrent working conditions, poverty, grotesque inequality and ecosystem collapse. All of this to the backdrop of intensifying climate change that is already destroying communities, displacing people from their homes, ruining livelihoods

3   David Harvey, *A Companion to Marx's Capital* (London & New York: Verso Books, 2010), p. 147.

4   Andreas Malm, *Fossil Capital: The Rise of Steam Power and the Roots of Global Warming* (London & New York: Verso Books, 2016), p. 298.

and killing innocent people. Mark Fisher's theory of capitalist realism accounts for the strength of ideology in making capitalism common sense despite all this. Inspired by the slogan 'it is easier to imagine the end of the world than it is to imagine the end of capitalism', Fisher defines capitalist realism as: 'the widespread sense that not only is capitalism the only viable political and economic system, but that it is now impossible to even *imagine* a coherent alternative.'[5] Fisher notes that climate crisis and environmentalist critiques offer the strongest rebuttal to capitalist realism: 'far from being the only viable political-economic system, capitalism is in fact primed to destroy the entire human environment.'[6] How does capitalism neutralise this seemingly irrefutable challenge to its ideological foundations? While hard-line conservatives are distracted by conspiratorial climate change denial funded by the fossil fuel industry, capital placates everyone else by incorporating ecological and climate concern into its own culture and marketing.

In a classic example, Fisher describes the environmentalism of the Disney/Pixar movie *Wall-E* (2008). The world is overcome by waste due to consumerism. Human civilisation migrates to spaceships, where overconsumption gets even worse, while robots clean Earth. Eventually humanity returns to Earth and rediscovers the joys of abundant planetary living under the same corporate regime responsible for the original crisis. My Dad took me and my brother to see the film when it came out. On the way out he asked us what we thought the message was. As long as he could hold our pre-adolescent attention, we briefly discussed the themes of waste and overconsumption. Years later, upon reading Fisher's discussion of the film, I gained a new understand-

---

5   Mark Fisher, *Capitalist Realism: Is there no alternative?* (London: Zero Books, 2009), p. 2.
6   Ibid., p. 18.

ing of the function of *Wall-E*. Fisher argues: 'the film performs our anti-capitalism for us, allowing us to continue to consume with impunity.'[7] The idea is that having spent an hour and forty minutes watching a film about consumerism and environmental collapse, we can feel righteous about our critical understanding of the imperfections of our system without actually doing anything to change it. Capitalist popular culture is not denying the system's internal problems (indeed, it profits from satirising them). It just denies that an alternative system is possible or desirable.

Popular concern for climate change is co-opted by corporations into their marketing and public relations campaigns. From 2020, meat substitute company Quorn put climate change at the centre of advertising campaigns. One ad said: 'We care about the world around us more than ever and we love our food. So Quorn Crispy Nuggets are a step in the right direction because they help us reduce our carbon footprint and they taste amazing.' Another advert is a YouTube collaboration with corporate partner Liverpool Football Club. Players Xherdan Shaqiri, Alex Oxlade-Chamberlain and Jordan Henderson awkwardly laugh through their corporate obligation obviously uncomfortable with the fact that none of them are close to being vegetarian. Henderson brings some respectability to proceedings with a dignified remark: 'To have everything in moderation, that's the best way.' Companies like Quorn don't have much interest in transforming the food system to eliminate workers' rights abuses, environmental degradation and emissions. For this corporation, individuals 'reducing their carbon footprint' and practicing moderation is conveniently synonymous with buying their product.

---

7   Ibid., p. 12.

# THE C-WORD (CAPITALISM)

The first time I encountered the climate justice marketing of Ben & Jerry's ice cream was in November 2015. I'd just arrived in Paris for the annual youth climate conference preceding the UN climate summit. On the wall of a Metro station was a large Ben & Jerry's poster comparing their ice cream to the planet: 'Quand c'est foundu, c'est foutu!' (When it's melted, it's fucked!), also promoting the climate march organised for the day before the summit. My first impression was of the hypocrisy of a dairy company advertising its support for climate justice while profiting from emissions-intensive animal agriculture. I later learned that although Ben Cohen and Jerry Greenfield are progressive activists, including funding Bernie Sanders' campaigns for President and campaigning to reduce corporate influence in politics, the company was bought by multi-national conglomerate Unilever in 2000. Supporting climate justice, migrants' rights, or Black Lives Matter is a just a brand for this subsidiary of Unilever. For consumers, buying Ben & Jerry's is a satisfying expression of progressive values, as if they're making some difference to the climate crisis. Unilever enhances the environmentalist reputation of its subsidiary and, most importantly, sells more products to make a profit. Capital is driven to turn everything into a commodity, including anti-capitalism.

Capitalism's evolution into its neoliberal stage was driven by the aims of consolidating capital's power over labour and securing new markets to profit from. Neoliberalism's headline ideological promise was reducing governmental interference in the economy. Actually existing neoliberalism has instead redirected state power towards imposing and upholding deregulation, privatisation, de-politicisation and austerity. These reforms began in the 1970s and 1980s amid a crisis of high inflation and high unemployment beginning with Augusto Pinochet's dictatorship in Chile, and the elections of Margaret Thatcher and Ronald Reagan in the UK and US. Another wave of reforms came under the guise of austerity after the 2008 financial crisis. Neoliberalism has emerged roughly in parallel to the intensification of climate crisis. Some think of this as an unfortunate accident of history, cruelly impeding our attempts to decarbonise (the effort to reduce carbon dioxide and other emissions across the economy), but neoliberalism was designed and (with the full force of the state) imposed to inject a few more decades of profitable life into capitalism as it faced existential crises of its own making. Neoliberalism is not the root cause of climate injustice, but its reforms have facilitated capital's continued profiting from it.

Deregulation has meant stripping away any constraints on capital's right to profit, most significantly for the climate, including fossil fuel companies and the banks that finance them to discover and extract fossil fuels. The privatisation of public assets has handed key sectors of industry to capital. Throughout the 1980s and 1990s, UK Prime Ministers Margaret Thatcher and John Major privatised large sections of the economy by selling state-owned utilities to private investors. This included Jaguar, British Telecom, British Aerospace, Britoil, British Gas, British Steel, British Petroleum, Rolls Royce, British Airways, British Coal and British Rail. What is striking about this list of transport, manufacturing and energy companies, is the role they might have played in a government-led energy transition had they not been stripped of democratic public control. With deregulation and privatisation comes de-politicisation. There's no need for democratic politics to interfere with the 'free market' as it maximises profits and neglects all else. With industrial strategy abandoned and finance capital supreme, the failure to decarbonise makes clear that whether capital or governments are in charge, the economy is political and prioritising profit has dire consequences. The austerity imposed after the 2008 financial crisis was as political a choice as any. In the UK, David Cameron's government presented austerity as necessary to balance the books amid bloated government debt and a public spending deficit. Government investment and spending was demonised as irresponsible and austerity was used as cover for finishing what Thatcher started: hollowing out the remains of the welfare state. Government subsidies for renewable energies and green homes (insulation and retrofitting new boilers) were scrapped. They also sold the UK's green investment bank, which it had launched to inject public money into green projects including energy efficiency and renewable energy.

Regardless of how little it was understood, climate change existed before neoliberalism. It will outlast it too. Returning to a pre-neoliberal capitalism by rolling back the last 50 years' reforms doesn't tackle the root of the problem. But by struggling for stronger regulation, expanding public ownership across the economy, democratising industrial strategy, and investing, we can push capitalism to its breaking point on a journey to climate justice.

A further pillar of capitalism's endurance is a series of governance and financial institutions existing to uphold capital's interests and reproduce the system. In the 1980s and 1990s, the World Bank and International Monetary Fund used 'structural adjustment' programs which attached conditions to introduce domestic neoliberal reforms to loans provided to Latin American and African countries experiencing economic crisis. Today, the United Nations manages the official multilateral climate negotiations through the United Nations Framework Convention on Climate Change (UNFCCC), producing trade agreements under the guise of climate action. Multinational banks provide financial support for capital projects, most prominently fossil fuel infrastructure, while governance institutions provide political support. The Trans-Adriatic Pipeline is a prime example, perversely designated as a 'Project of Common Interest' by the European Commission. The only 'common interest' about it is between the fossil fuel companies profiting from the gas they'll transport from Azerbaijan, through Greece and Albania, to Southern Italy. The pipeline is opposed by locals in Puglia in Italy over its climate impacts and threat to local farming livelihoods. A campaign for banks to defund the expansion failed with the European Investment Bank and European Bank for Reconstruction and Development providing loans of €1.5 billion and €1 billion respectively. Banks made this unpopular, socially and ecologically devastating pipeline financially viable, while the EU made it politically legitimate.

We can often discuss 'capital' as unified in its interest of maximising profit, but we should also understand that capital takes different forms (by economic sector) which are often in competition with each other. Some of these fractions of capital, like fossil capital and agribusiness, are self-evidently existentially implicated in the climate crisis. Their core business model directly produces emissions. Other fractions, like finance, could conceivably continue to profit in a post-fossil-fuel capitalism. A stable climate could secure decades or centuries of more profitability for finance and other fractions. So why is finance capital one of the chief supporters of fossil capital's expansion? Why does real estate not do its part and retrofit property? Why does the technology industry not invest in transforming its emissions-intensive infrastructure? These long-term divergences in planetary interest are subordinated to the strength of short-term drive to maximise profits by investing in the very fossil fuel infrastructure that will render the planet eventually incompatible with any economic activity. This logic produces a powerful class solidarity between fractions of capital, collaborating to uphold each other and mutually sustain the capitalist system they depend on. There will be no alignment away from profitable fossil fuels by capitalism, even as it expands the renewable energies market. As long as profit is supreme, climate change will get much worse.

## Fossil capital

Fossil fuel extraction is woven into the DNA of capitalism, emerging as the system's energy source of choice due to its ability to satisfy capital's drive for profit. In *Fossil Capital*, Andreas Malm identifies a key moment in capitalism's development, in eighteenth century Britain, with the transition from water power to coal power. The foundational relationship between the exploitation of labour and nature comes to the fore again. The transition occurs for capital to most efficiently exploit workers who were increasingly concentrated in urban areas. Though water power had its advantages, it was less mobile than fossil fuels, which freed industry from the constraints of proximity to water. Until then, capital had continued to accumulate by accessing more and more fertile land, before this inevitably led to expansion into inferior, less profitable land.[8] Fossil fuels allowed capital to break free from the stagnation of finite space. The rise of fossil fuels as the global economy's primary energy source was not a politically neutral accident of history, but an essential event in capitalism's development for it to survive.

Today, fossil capital is managed by a mix of private and state-owned companies: the fossil fuel industry. Some of these companies (like BP and Shell) are household names. There

---

8   Malm, *Fossil Capital*, pp. 22, 124, 146.

are hundreds more with less name recognition also continuously exploring for and extracting oil, gas and coal reserves too. To keep global warming below even 2°C, over 80% of these reserves must stay in the ground. However, the business model of these companies is to constantly expand their reserves and then extract until there's no more. The industry organises hard to protect that business model by lobbying politicians, campaigning against climate-friendly legislation and investing in their own corporate image as protectors of the future. In 2018, BP and others funded the campaign against 'Proposition 112' to restrict fracking (also known as hydraulic fracturing: a method of injecting a pressurised liquid into wells to release gas) in Colorado. Weeks later after Prop 112 was defeated, BP launched an ad campaign promoting its investments in reducing emissions through 'bioenergy'.[9]

Activists and climate scientists alike have called for fossil fuel executives to be tried for crimes against humanity, drawing parallels between their actions and the classical definition: a widespread or systematic attack directed against any civilian population, with knowledge of the attack, including murder and extermination.[10] There's power to this argument. Fossil fuel executives make for a clear enemy, having stewarded the companies most responsible for emissions in full knowledge of their role in climate injustices. But there's a danger of making the climate crisis about the executives, and not about the system they operate within. The CEOs of Shell or BP could step down

9   Sandra Laville & David Pegg, 'Fossil fuel firms' social media fightback against climate action', *Guardian*, https://theguardian.com/environment/2019/oct/10/fossil-fuel-firms-social-media-fightback-against-climate-action (last accessed 15 June 2020).

10   Kate Aronoff, Alyssa Battistoni, Daniel Adana Cohen & Thea Riofrancos, *A Planet to Win: Why We Need a Green New Deal* (London & New York: Verso Books).

to pursue a more ethical career, but their place would be seamlessly filled by someone else. Shareholders can sell their stocks in protest, but they will be bought up by another speculator. The companies and their executives have little choice in whether they continue extracting. They are structurally unable to transition from fossil fuels to renewables of their own accord. As long as there are fossil fuels to be discovered and extracted, capital's drive to accumulate means they will be – or else everyone loses lots of money.

## Fossil states

There is only one political form presently capable of dismantling the fossil fuel industry on the timescale that the climate crisis commands: the state. Given the industry's multinational scale and capital's mobility, states will have to collaborate across borders. If states have the capacity to rein in fossil capital, why haven't they begun to? That the state is a functionally useful political form doesn't mean it is essentially good. Although radical climate movements have emphasised the culpability of the private fossil fuel companies (for example, through divestment campaigns for investors to sell shares in fossil fuel companies), operating within a capitalist context, nation states actively uphold fossil capital, including themselves acting as fossil fuel companies. The idea of a 'petrostate' refers to a country with a heavy economic reliance on oil and gas. This might bring to mind a kingdom in the Middle East or a Latin American country trying to keep its geopolitical distance from the US. Fossil fuels may form a proportionately large part of these countries' economies, but the incestuous relationship between fossil capital and state power implicates states of all geographies, sizes and statures.

The United States spent much of the 1990s and 2000s waging war in the Middle East to secure access to the region's oil supply. Domestically, the federal government has provided billions of dollars in subsidies for coal, oil and gas, including a booming fracking industry. In 2016 and 2017, Indigenous nations in North America converged at the Standing Rock Reservation to resist the construction of the Dakota Access Pipeline to protect water, land and sacred sites. The government's response demonstrated a willingness to use violent military tactics (as police fired water cannons and rubber at protectors) to secure the domestic interests of fossil capital. The Canadian government provides political, financial and diplomatic support for its large mining industry, which profits from extraction around the world. In May 2018, Prime Minister Justin Trudeau announced his government would nationalise the TransMountain oil sands pipeline to ensure its expansion was built in the face of mass resistance. In the UK, the government provides financial support for North Sea oil and gas exploration and spent millions on policing to protect a nascent fracking industry while protestors waged a protracted direct-action campaign against it.

China is sometimes praised for its decarbonisation efforts. In 2020, China announced its ambition to reach net-zero emissions (a state of carbon neutrality, achieved by reducing emissions and offsetting what's left with carbon removal) by 2060. It is also heralded as the world's largest producer of renewable energy or for having the most electric buses in the world. At the same time, China operates its own state oil and gas company, China National Petroleum Corporation, with operations around the world. Its state-owned banks, Bank of China and Industrial and Commercial Bank of China, are (along with publicly traded China Reconstruction Bank) the world's biggest financiers of coal mining and coal power globally. China's multi-trillion-dollar global development strategy, the Belt & Road initiative, spanning Asia, Africa and Europe, has earmarked over $36 billion of investment for coal power internationally.[11] China's energy strategy reflects that of capital in the age of climate crisis: new renewable energy markets at home, while exporting fossil fuel expansion abroad.

While fossil fuels are just one part of imperial states' economic domination, smaller Gulf states are more wholly dependent on their fossil fuel reserves. The kingdoms of Bahrain, Kuwait, Qatar, Oman, United Arab Emirates and Saudi Arabia operate state-owned oil companies respectively accounting for significant proportions of GDP and export revenue. Saudi Arabia, the biggest among them, has used this wealth to imitate the imperialism of the kingdom's Western backers, leading a coalition of forces in a brutal bombing campaign of Yemen. Others, like UAE and Qatar, have extended their power in the cultural sphere, buying football clubs Manchester City and Paris Saint Germain, while Qatar was awarded the 2022 FIFA World Cup.

Naomi Klein uses the examples of the Soviet Union and Venezuela as 'a reminder that there is nothing inherently green about self-defined socialism.'[12] The economic development of both was reliant on the extraction of domestic coal and/or oil reserves, but Venezuela is not a fully-fledged socialist society. It's a revolution in motion, operating within a context of global capitalism, and of blockade and sabotage by the US government and US capital. Oil has long been central to Venezuela's

---

11   Institute for Energy Economics and Financial Analysis, 'IEEFA China: Lender of last resort for coal plants', https://ieefa.org/ieefa-china-lender-of-last-resort-for-coal-plants/ (last accessed 16 December 2020).

12   Naomi Klein, 'Capitalism Killed Our Climate Momentum, Not "Human Nature"', *Intercept*, https://theintercept.com/2018/08/03/climate-change-new-york-times-magazine/ (last accessed 27 July 2020).

economy. George Ciccariello-Maher describes the 'decade of oil-fuelled growth' which preceded 'a crisis since at least 1983, when the price of oil tanked and the currency devalued sharply, instantly making people's wages and the money in their pockets worth much less.[13] Responding to neoliberal reforms imposed on Venezuela amid oil crises, a revolt led by socialist Hugo Chávez eventually led to his election as President in 1998. After defeating a coup attempt in 2002, Chávez gained full control of state oil company PDVSA and began spending on a series of social welfare programs, including food, housing, education, healthcare, land reform and indigenous rights. Today, nation states in the Global South have no real choice regarding fossil fuel extraction. Either natural resources are open to Western capital to plunder through private markets, in exchange for capitalist development (fuelling poverty and inequality), or the state controls domestic oil wealth to fund the population's basic needs. Ciccariello-Maher describes the widely reported political instability, inflation, and food shortages that followed Chávez's death and succession by Nicholás Maduro: 'For this, the collapse in oil prices is partly to blame, but more so the longstanding failure to break oil dependency by stimulating domestic production. This is a century-old contradiction, and the problem is in the system, not in the government, as the opposition would argue.'[14]

If we are to seriously confront the climate crisis, we have to start with this understanding that capitalism is both the root cause and biggest barrier to change. It is capital's insatiable drive for more and more accumulation by exploiting workers and the environment which keeps us on this path towards even greater catastrophe. The profits of capital are upheld by ideo-

---

13    George Ciccarellio-Maher, *Building the Commune: Radical Democracy in Venezuela* (London & New York: Verso Books, 2016), p. 3.

14    Ciccarellio-Maher, *Building the Commune*, p. 130.

logical dominance in media and culture, neoliberal reforms, international political and financial institutions, and powerful class solidarity between capital itself. The priority of capitalism to pursue profits above all else means that nation states remain dependent on fossil fuel extraction and fossil capital has no choice but to carry on extracting. Transforming this system won't be easy, but if we are committed to justice, it is absolutely necessary.

# Chapter 2

# Justice or bust

In 2020, Storms Ciara and Dennis gave England and Wales the wettest February since records began. Flooding spanned the country and the news showed people wading through several feet of murky brown water filling their homes and businesses. A Sky News interview with Vic Haddock, an affected Shropshire homeowner, went viral: 'I'm a staunch supporter of Boris Johnson.' It was just two months since Vic voted for Johnson in the 2019 general election. 'Now I've supported him. Come on Boris, come and support me.' Vic's wearing a Wolverhampton Wanderers beanie and there's a canoe floating behind him, where the road should be. He says he accepts that he's bought a house by the river: 'I've bought this of my own accord. I expect to get my feet wet. But I don't expect this.' When the interviewer said the government were offering up to £500 in support, he laughed. '£500! What's that going to do? There's £500 worth of damage in the fridges and the freezer alone.' His face was then straight when asked if he had insurance. 'No. Can't get insurance. It's just ordinary working people down here, and they come up with extortionate rates. Seven grand a year or something like that.' In any case, the insurers won't provide any cover for flooding. They know they won't make a profit, so they don't insure the risk.

For most people, there is no safety net for when their homes are hit with thousands of pounds worth of damage. For those able to afford it, owning a home is supposed to be a source of financial security in an increasingly precarious economy. When floods hit, that security is taken away by profiteering insurance firms and neglectful governments. This is climate injustice. The impacts of climate change are distributed unfairly so that those who have contributed the least to the crisis are hit the hardest, while those who have actively driven climate change experience the least harm and profit the most.

The injustices of flooding are not just felt in the UK, but even more harshly across the world. Later in 2020, at one point in July, over a third of Bangladesh's land was under water. The previous July, Al Jazeera reported: 'Rain-swollen rivers in Bangladesh have broken through several embankments, submerging dozens of villages, destroying tens of thousands of homes and displacing nearly 200,000 people.'[1] By the end of the month, over 100 people had died. Flooding on this scale will now be an annual occurrence for Bangladesh. There are too many places affected by flooding in 2020 alone to list them all, but from Kenya to Afghanistan to Vietnam, deadly floods will only become more frequent and severe. Some of the most densely populated cities in the world are at the greatest risk. The people living there have contributed nothing to global emissions. Their governments have contributed relatively little too. Yet these countries remain at the highest risk, unable to secure themselves against rising sea levels and extreme weather.

In *The Uninhabitable Earth*, journalist David Wallace-Wells writes that five or six degrees of global average temperature rises will mean 'whole parts of the globe would be literally unsur-

---

1  'Worst floods in years "submerge" Bangladesh villages', *Al Jazeera*, https://aljazeera.com/news/2019/07/worst-floods-years-submerge-bangladesh-villages-1907 19083053518.html (last accessed 14 May 2020).

vivable for humans.'[2] Relatively mild climates like New York will become hotter than anywhere found in the world today. Labouring in the summer will become impossible in many places. Wallace-Wells says that this scenario is unlikely before 2100, but '255,000 workers are expected to die globally from direct heat-effects' by 2050.[3] The turn of the next century is now within the lifetimes of many born today and workers are already feeling the injustices of planetary heating. Hundreds of migrant labourers working in Qatar in preparation for the 2022 FIFA World Cup are reported to have died in the heat. The COVID-19 pandemic has shown what bosses are prepared to put workers through during a crisis. London buses and underground trains packed full of low-paid workers forced to commute at the height of the pandemic will be mirrored by images of low-paid labourers toiling in the deadly heat of intense climate change. Their choice will be to work to death for a meagre wage or to starve. As wealthy and professional classes have greater freedom to work from air-conditioned homes during pandemics or heatwaves, manual workers remain fatally exposed to the elements.

## An imperial overheating

The Middle East will be one of the region's most intensely affected by global heating. It also happens to have been politically destabilised by decades of Western imperialism with war, occupation and military intervention in Iraq, Afghanistan, Libya, Syria and Yemen as notable examples. Western ambitions for geopolitical control in the region have often been tied to access to natural resources, including oil and the creation of

2 David Wallace-Wells, *The Uninhabitable Earth* (London: Penguin Books, 2019), p. 39.
3 Wallace-Wells, *Uninhabitable Earth*, p. 48.

new markets for capital to operate in. The consequences have been millions of deaths and the political conditions which gave rise to ISIS across the region and a slave trade in Libya. Heating will continue to add to the death count and erode ways of living. Journalist Remona Aly has called on Muslims to join the climate movement because of the threat that planetary heating poses to hajj – a pilgrimage made by two million Muslims every year to Mecca in Saudi Arabia. The pilgrimage is mandatory at least once in the lives of Muslims but will become physically impossible due to temperature rises. 'Heat and humidity levels during hajj will exceed the extreme danger threshold 20% of the time from 2045 and 2053.'[4]

It's not just Middle Eastern countries on the receiving end of resource wars and occupation. With Western support, Indonesia has occupied West Papua (the western half of the island of New Guinea) since 1963 when the Dutch withdrew from the colony. The Indonesian government promotes extraction of natural resources, including oil and gas, in the name of 'development'.[5] Indonesian security forces facilitate the violations of Indigenous land rights by multinational corporations (such as, BP, Rio Tinto and BHP Billiton) so they can profit from extraction. 'West Papua contains one of the world's largest tropical rainforests – covering around 30 million hectares – but this huge area of biodiversity is under threat.'[6] Through occupation and colonial

---

4   Remona Aly, 'With hajj under threat, it's time Muslims joined the climate movement', *Guardian*, https://theguardian.com/commentisfree/2019/aug/30/hajj-muslims-climate-movement-global-heating-pilgrims (last accessed 14 May 2020).

5   Jason MacLeod, *Merdeka and the Morning Star: civil resistance in West Papua* (Queensland: University of Queensland Press, 2015), p. 60.

6   Connor Woodman, 'SACRIFICE ZONE: BP, FREEPORT AND THE WEST PAPUAN INDEPENDENCE STRUGGLE', *New Internationalist*, https://newint.org/features/2017/05/01/sacrifice-zone-west-papuan-independence-struggle (last accessed 14 May 2020).

extraction, indigenous West Papuans are displaced from their land while deforestation contributes to worsening climate change. On 1 December 2020, the United Liberation Movement for West Papua (ULMWP) began to form a new provisional government, stepping up the struggle against colonisation, with the aim of mobilising West Papuan people to achieve a referendum on independence, 'after which it will take control of the territory and organise democratic elections.' Benny Wenda, interim President of the Provisional Government, and chairman of the ULMWP, declared: 'As laid out in our Provisional Constitution, a future Republic of West Papua will be the world's first Green State, and a beacon of human rights – the opposite of decades of bloody Indonesian colonisation.' While capitalist states commit the most egregious human and ecological crimes in pursuit of profit, the ULMWP resists the occupation with a vision of freedom, indigenous stewardship of the environment, and climate justice through solidarity.

## Never trust a COP

Since the ratification of the United Nations Framework Convention on Climate Change (UNFCCC) in 1994, the UN has been responsible for coordinating the international effort towards the 'stabilization of greenhouse gas concentrations in the atmosphere at a level that would prevent dangerous anthropogenic [human-caused] interference with the climate system'.[7] Under the guise of international climate action, the annual Conference of the Parties (the parties being member states), aka COP, have acted as key moments in reproducing global injustices and safe-

---

7   United Nations Framework Convention on Climate Change, Article 2, https://unfccc.int/resource/docs/convkp/conveng.pdf (last accessed March 2021).

guarding capitalism against retribution or transformation. Joel
Wainwright and Geoff Mann argue that COP treats capitalism
'not as a question, but as the solution to climate change.'[8] The
Kyoto Protocol was the UNFCCC's first landmark, adopted in
1997 and coming into force in 2005, though it was not ratified
by the US under George W. Bush's Presidency, and Canada later
withdrew under Stephen Harper's government as it led the
drive to extract Alberta's oil sands. Under the treaty's principles, developed countries would take on proportionally greater
obligations to reduce emissions owing to historic responsibility.
However, its mechanisms introduced carbon markets, letting
large emitters off the hook by allowing them to trade their obligations to other countries.[9] Put simply, wealthier countries
with more emissions could pay poorer countries to make extra
emissions reductions on their behalf.

COP15, hosted in Copenhagen in 2009, ended in failure.
The US, European and BRIC countries (Brazil, Russia, India,
China) constructed their own deal behind closed doors, which
was rejected by excluded countries from the Global South. The
Parties could not agree anything beyond a statement (not even
passed unanimously) supporting actions to reduce emissions,
without targets or legal obligations. Six years later, at COP21
in Paris, world leaders achieved what they could not in Copenhagen. The Paris Agreement was signed with an ambition to
keep global average temperature rises below 2°C, or prefera-
bly 1.5°C. Celebrated though it is, the Paris Agreement strips
away any pretence of global justice. 'The Paris Agreement does
not separate party-states into groups with different commit-

8   Joel Wainwright & Geoff Mann, *Climate Leviathan: A Political Theory of Our Planetary Future* (London & New York: Verso Books, 2018), Chapter 2, Section III.

9   UNFCCC, Emissions Trading, https://unfccc.int/process/the-kyoto-protocol/
mechanisms/emissions-trading (last accessed 23 December 2020).

ments based on wealth or income, unlike the Kyoto Protocol of 1997.[10] Decarbonisation is based on 'nationally determined contributions' (NDCs) where countries decide their own emissions reductions targets, with no legal mechanism of enforcement. High-emitters have a free pass to continue with business-as-usual. If all the NDCs were to be achieved, then global average temperature rises would only be kept to around 3.2°C by the end of the century.[11]

A popular refrain among global climate justice groups is '1.5° to stay alive'. Charting the path to runaway climate change, the Paris Agreement is a death sentence to many around the world. By abstracting questions of climate change to scientific jargon and marginal global average temperature increases, the UNFCCC rejects the concept of justice, including ignoring the existence of Indigenous peoples. Tom B.K. Goldtooth, the Executive Director of the Indigenous Environmental Network, responded to the agreement saying it 'treats Nature as capital with no real nor effective safeguard mechanisms that could guarantee the prevention of land grabs and the protection of the rights of Indigenous peoples.'[12] The UNFCCC acts as cover for states and corporations to claim a commitment to climate action while creating new capitalist markets and protecting the geopolitical domination of those most responsible for climate injustices.

---

10  Wainwright & Mann, *Climate Leviathan*, Chapter 2, Section III.

11  UN Environment Program, Emissions Gap Report 2019, p. 27, https://wedocs.unep.org/bitstream/handle/20.500.11822/30797/EGR2019.pdf?sequence=1&isAllowed=y (last accessed 18 May 2020).

12  Indigenous Environmental Network, 'Indigenous Environmental Network: REDD "must be immediately canceled"; https://redd-monitor.org/2016/05/24/indigenous-environmental-network-redd-must-be-immediately-canceled/ (last accessed 18 May 2020).

## A racist crisis

We're not saying that climate change affects only black people. However, it is communities in the global south that bear the brunt of the consequences of climate change, whether physical – floods, desertification, increased water scarcity and tornadoes – or political: conflict and racist borders. While a tiny elite can fly to and from London City airport, sometimes as a daily commute, this year [2016] alone 3,176 migrants have died or gone missing in the Mediterranean, trying to reach safety on the shores of Europe.[13]

In 2016, Black Lives Matter in the UK (BLMUK) launched with a focus on combatting environmental racism. They borrowed the direct-action tactics of the environmental movement to disrupt major London airports, blocking a main traffic entrance to Heathrow and occupying the runway of London City Airport. BLMUK highlighted the environmental harms felt disproportionately by working-class black Londoners in the communities surrounding the airports used predominantly by the rich. They drew the links to the racialised distribution of climate impacts globally, stating in a video: 'Seven out of ten of the countries most affected by climate change are in sub-Saharan Africa.' Those very countries are side-lined in the UNFCCC negotiations which claim to represent international climate action. At COP15 in Copenhagen, Lumumba Di-Aping, Sudanese diplomat and chief negotiator for the G77 group of developing countries,

---

13  Alexandra Wanjiku Kelbert, 'Climate change is a racist crisis: that's why Black Lives Matter closed an airport', *Guardian*, https://theguardian.com/comment isfree/2016/sep/06/climate-change-racist-crisis-london-city-airport-black-lives-matter (last accessed 7 February 2021).

described the deal on offer as a 'suicide pact' for Africa.[14] From the boroughs of London to the geographic distribution of extreme weather, to international climate diplomacy, racism is central to climate injustice.

The racism of the climate crisis is a product of the foundational racism of capitalism and colonialism. A disregard for black lives was necessary for capitalism to be built on the unpaid labour of a hyper-exploitable racialised under-class, from slavery at its dawn to the disproportionately incarcerated black people in today's prison industrial complex. For example, when slavery was abolished in the US, capital's need for free labour to maintain profitability didn't go away. Black Americans have since been disproportionately incarcerated in the bloated prison system. As of 2019, black men are 5.7 times more likely to be imprisoned compared to white men.[15] Earning wages of less than $1 per hour (in Texas, prisoners are not paid for work at all), 7% of state prisoners and 18% of federal prisoners are employed by private companies, usually in farming work or performing repetitive tasks producing consumer goods. This produces around $2 billion of value every year.[16] Prisoner Kevin Rashid Johnson wrote in the *Guardian*: 'I see prison labor as slave labor that still exists in the United States in 2018. In fact, slavery never ended in this country.'[17] In California, where fatal wildfires are

---

14   Leon Sealey-Huggins, 'The climate crisis is a racist crisis: structural racism, inequality and climate change', in *The Fire Now: Anti-Racist Scholarship in Times of Explicit Racial Violence*, ed. Johnson, Azeezat, Joseph-Salisbury, Remi and Kamunge, Beth (London: Zed Books, 2018), p. 104.

15   US Bureau of Justice Statistics, Prisoners in 2019, https://bjs.gov/content/pub/pdf/p19.pdf, p. 16 (last accessed 25 March 2021).

16   NAACP, Criminal Justice Fact Sheet, https://naacp.org/criminal-justice-fact-sheet/ (last accessed 24 December 2020).

17   Kevin Rashid Johnson, 'Prison labor is modern slavery. I've been sent to solitary for speaking out', *Guardian*, https://theguardian.com/commentisfree/2018/aug/23/prisoner-speak-out-american-slave-labor-strike (last accessed 24 December 2020).

now an annual event, the huge gaps in the state's fire service left by decades of privatisation are filled by 'volunteer' (unpaid) prison labour.[18] Yet again, it is those who have contributed least to climate change and are exploited most by capital that are put in harm's way when its impacts hit hard.

Colonialism was crucial to imposing capitalism around the world, forging new sites of extraction, exploitation and markets for capital to profit from. Foundational to colonialism and Empire was the genocide of Indigenous peoples and occupation of their territory. Leon Sealey-Huggins writes, 'To be clear, a specific consequence of the legacies of imperialism and colonialism, and contemporary forms of neoliberal capitalism, is the globally unequal distribution of wealth that leaves many Caribbean countries without the resources necessary to adequately respond to climate change.'[19] The racism of the climate crisis is the racism of capitalism, imperialism and colonialism, mediated through the great ecological consequences of capital's dominance. Countries in the Caribbean and across the Global South are especially vulnerable to climate change, but Sealey-Huggins argues that simple geography is not the reason. Peoples in these countries experience climate impacts most harshly because of the structural racism of capitalism and colonialism.[20] Though most now have formal independence from Empire, colonial relations continue to force these 'vulnerable' countries into bonds of debt, limiting their ability to invest in mitigation and adaptation. They remain dependent on economic

18   Beryl Lipton, 'Inmates make up nearly a third of California's firefighting force', *Muckrock*, https://muckrock.com/news/archives/2018/aug/14/ca-firefighters-prison-labor/ (last accessed 24 December 2020).

19   Leon Sealey-Huggins, '1.5°C to stay alive: climate change, imperialism and justice for the Caribbean', *Third World Quarterly*, 38 (2018), p. 2451.

20   Sealey-Huggins, 'The climate crisis is a racist crisis', p. 103.

sectors most susceptible to climate shocks such as fishing, agriculture and tourism.[21] These inequalities within and between nations experiencing climate change are not natural. They are reproduced by capitalism's political and financial institutions to maintain the profitability of the system, even as Empire formally ends.

## Just transition

There are a few key elements to decarbonisation. The first is expanding the production of renewable energy (solar, wind, tidal, geothermal and hydropower). For renewables to be viable, greater production requires concurrent investment in battery technologies to expand storage capacity. The second element is ending the extraction and burning of fossil fuels. It doesn't matter to the climate how much renewable energy we produce if we don't also end carbon emissions by dismantling the fossil fuel industry. Shifting entirely from fossil fuels to renewables is the energy transition. The third element is the 30% of emissions which come from other sources: deforestation (which causes stored carbon to be released), industrial agriculture, and cement, steel and plastic production (which cause significant methane emissions).[22] Rather than transition per se, these other emissions sources require a transformation of certain economic sectors. Agriculture and industrial production do not need to end like fossil fuels, but they need to change to be less polluting and energy intensive. We need to reduce energy usage across

---

21   Sealey-Huggins, '1.5 to stay alive', p. 2445.

22   Jason Hickel, 'Clean energy won't save us – only a new economic system can', *Guardian*, https://theguardian.com/global-development-professionals-network/2016/jul/15/clean-energy-wont-save-us-economic-system-can (last accessed 18 May 2020).

the economy, concurrent with decarbonisation. If economic production expands, then so does demand for energy. This makes decarbonisation more challenging. We'd require more renewables infrastructure, which in turn means more mining of mineral resources and more carbon emitted for their production (in the time it takes to fully decarbonise). It requires the generation of more storage capacity and altogether more time to fully decarbonise.

Together, the energy transition and industrial transformations present serious social threats, as well as the obvious economic and ecological opportunities. Capital has traditionally used moments of economic upheaval to cut jobs, erode trade union power and impose austerity on working people. The legacy of Margaret Thatcher's assault on Britain's mining communities and trade unions in the 1980s is still felt in towns and communities plagued by structural unemployment. Today, most workers in the UK's oil and gas industry already have no employment rights due to their status as contractors. With more busts than booms in the industry's business cycle, their work is precarious too. Those workers' shared anxiety is that the energy transition will be used to further erode pay and conditions, or that jobs will disappear entirely as they're outsourced to places where labour costs are cheaper. For workers in the high-carbon aviation industry, the COVID-19 pandemic has shown what is to come. With flights grounded, thousands of jobs were cut and many of those that survived were fired and rehired on contracts with worse conditions. As climate change brings more shocks to high-carbon industries, the companies will continue to assault workers' pay and conditions. Destroying workers and their families' livelihoods must not be the price we pay for decarbonisation. These workers have been exploited by the companies profiting from climate injustice and are now at risk of being among its greatest victims. There is an alternative though: a just transition.

The principles of just transition emerged from the trade union movement to promote an energy transition which protects workers and communities most at risk. A just transition stipulates that all workers retain an equivalent job with the same terms and conditions, pay and trade union rights. By introducing forms of economic democracy, workers and communities could use their considerable expertise of their local area and the energy sector to decide which new green industries they will transition to. It could ultimately end with workers managing new public companies in new green industries, rather than continue to be exploited in insecure work by profiteering private companies. If the energy transition is left to existing fossil fuel companies, then it simply won't happen on the timeline required. It certainly won't incorporate any of workers' demands for justice. Though investment in renewables has begun, the allure of fossil fuel profits is too great for private companies to hammer the nail in the coffin of their prime source of revenue. If the same companies begin to operate renewable energy production too, workers can just expect the same terrible treatment. A just transition therefore requires governments to manage the fossil fuel industry's decline while guaranteeing equivalent green jobs in new public companies.

## Just adaptation

Even if we were to decarbonise on the quickest timetable possible, historic emissions have already locked in devastating climate impacts for years to come: more flooding, heat, drought and extreme weather. The longer decarbonisation takes, the more frequent and severe they will be, spreading to new, unexpected places. Equally important as mitigating climate change, then, is adapting to it.

Adaptation is sometimes framed as an alternative to mitigation, as if we should give up on arresting climate change and find ways to live with it instead. Such a position is either promoted by those looking to avoid confronting fossil capital and transform the economy, or by pessimists who don't believe we can. Jem Bendell is the author of a viral paper, 'Deep Adaptation: A Map for Navigating Climate Tragedy', in which he assumes a dire climate scenario and calls for deep adaptation to climate tragedy with three prongs. He argues for adaptation through resilience (psychological, as well as material), relinquishment (of elements of modern civilisation, including inhabiting coastlines, certain industry and consumption habits) and restoration (of older ways of life including re-wilding, seasonal diets and non-electronic forms of play).[23] Bendell's vision of adaptation is an austere primitivism that would be as punishing to those blameless for climate change as climate impacts themselves. Adaptation must be neither a lifeline for fossil capital nor a dystopian retreat from civilisation. Instead, a just form of adaptation can complement rapid decarbonisation with expanded state capacity and significant investment to secure our societies and economies against climate shocks. Emergency services must be fully funded and state run. We need to be confident that fire brigades are resourced to tackle the next wildfire and that hospitals can provide healthcare free at the point of use to anyone who needs it in a time of crisis. COVID-19 has demonstrated the fatal consequences of underfunded or privatised health services, where decades of cost cutting left insufficient capacity to respond to moments of

---

23   Jem Bendell, 'Deep Adaptation: A Map for Navigating Climate Tragedy', IFLAS Occasional Paper 2, https://lifeworth.com/deepadaptation.pdf (last accessed 18 May 2020).

crisis. We need to invest in upgrading buildings and other infrastructure to adapt to heating, flooding and extreme weather. At-risk areas require robust flood defences. This is expensive stuff, which the market won't deliver of its own accord unless it's profitable. That's why we need a democratic state to manage investment where it is needed.

As well as adapting our built environment, a just adaptation must include investing in resilient ecologies with new forms of agroecology, rewilding and reforestation to maximise the 'natural' forms of carbon sequestration (the removal of carbon from the atmosphere) that industrialisation has diminished. In some contexts, most crucial might be guaranteeing and expanding land rights for Indigenous peoples who have proven the most effective stewards of ecologies using traditional knowledges and relationships with nature. Indigenous peoples manage 25% of the world's land and 80% of biodiversity, despite representing less than 5% of total human population.[24] In the Brazilian Amazon, they are under attack from loggers and miners who were given free rein to seize Indigenous land by far-right President Jair Bolsanaro. For all the talk of technical mitigation and adaptation measures, ending the displacement of Indigenous peoples may be both the most instrumentally effective and morally just thing to do.

The painful reality is that some climate shocks will overwhelm even the most robust infrastructure, well-funded emergency services and restored biodiversity. To adapt to these scenarios, risk should be pooled collectively with insurance provided as a basic right. Private insurance companies have shown they will

---

24 Gleb Raygorodetsky, 'Indigenous peoples defend Earth's biodiversity – but they're in danger, National Geographic', https://nationalgeographic.com/environment/2018/11/can-indigenous-land-stewardship-protect-biodiversity-/ (last accessed 24 December 2020).'

not cover flooding or other disasters where they stand to lose too much. With state insurance, we can be sure that nobody loses out financially due to loss of property, livelihoods or life. Just adaptation would see this principle extended to the global level, with the international community collectively pooling climate risk, and with the most historically responsible countries and companies bearing the strongest obligations to pay in.

## Reparations

We can mitigate future emissions and adapt to what is unavoid-ably coming, but there are many who have already experienced irreversible harms and injustice. Some have paid for the profits of fossil capital with their lives. Climate justice means going as far as possible to repair those past or ongoing harms. Tradition-ally, reparations have been called for by African and Caribbean diaspora, colonised peoples, indigenous peoples, and immi-grants-rights activists for injustices including colonisation, occupation, displacement, slavery and genocide.[25] Students of history may also be familiar with the requirement of Germany and other defeated powers to pay financial reparations in the aftermaths of the First and Second World Wars for the costs they had caused. How then do we apply the concept of reparations to climate injustice? Given our monetary economy, financial transfers from those responsible should form part of the picture. Countries most responsible for emissions and the global impo-sition of fossil capitalism should pay for loss and damage arising from climate impacts and to support developing countries in their economic transitions. Companies that have profited

---

25   Raj Patel & Jason W. Moore, *A History of the World in Seven Cheap Things: A Guide to Capitalism, Nature and the Future of the Planet* (London & New York: Verso Books, 2018), p. 42.

from displacing communities from their land to make way for extractive projects should compensate for that injustice and to finance rebuilding those communities. As needed as financial reparations are, there is not enough money in the world to pay for what has already been lost to fossil capitalism and climate colonialism, let alone what is to come: the stolen life, land, livelihoods, communities and cultures. We need systemic and redistributive reparations beyond monetary terms.

Climate reparations should not just redress the unfair distribution of harms, but the unfair distribution of political power foundational to climate injustice. In the international arena, we need to reorganise the institutional architecture constructed over the last century to cement the dominance of the Global North. With a new set of governance and financial institutions, power between nations can be distributed more equally with climate processes led by countries worst impacted and financial bodies led by those who need support most. Domestically, the same political reckoning must occur where those hit hardest by climate impacts – the poorest, the working-class, women and Indigenous peoples – are guaranteed access to the resources and institutions necessary to influence decision-making. There's no point transforming politics if we don't transform the economy too. Undergirding political power is ownership of property and resources. The theft and misuse of mineral resources, fossil fuels and land, and the mismanagement of industry, all in the pursuit of profit, are at the root of climate injustice. Reparations for this must begin by returning natural resources, land and industry to public ownership so they can be democratically managed in the interests of present and future generations. For those who have lost so much (or have so much still to lose) from climate change, returning ownership of the economy is the beginning of reparations for climate justice by giving those affected the power to determine their own political, economic and ecological futures.

Just transition, just adaptation and systemic climate reparations cannot comfortably coexist with capitalism. The imperative for capital to accumulate at all costs, and the centrality of profit to all economic activity, means that our current economic system offers no incentive for capital and capitalist states to transition from fossil fuels to renewables, let alone protect workers' rights at the same time. There is no incentive for governments or corporations to provide monetary or redistributive reparations to contribute to reorganising power relations. Our reality, however, is that we must struggle for climate justice from within a changing climate and within capitalism as our starting point. Our task is to build this new economy from within the old. Every movement, campaign, resistance and mobilisation should be oriented towards this horizon as the linchpin of climate justice. By building and exercising collective power, we can create new economic logics – based on justice, solidarity and public ownership. By capturing state power, we can supplant capitalist logics by planning a just transition, just adaptation and reparations. As the protest chant goes: 'climate justice now!' It's up to us to create the conditions for it today.

# Chapter 3
# Climate Action, Ltd

12 December 2015. The Arc de Triomphe loomed over the many thousands converging in central Paris to defy the French government's state of emergency and protest COP21. Red was the theme of clothing, banners and the giant reams of cloth symbolising the 'red lines' the movement was drawing: 'the minimal necessities for a just and liveable planet.'[1] No more relying on politicians, governments or UN processes. Mainstream climate organisations and grassroots groups pledged to take the crisis into their own hands. That was the promotional messaging for the demonstration. On the ground, its tone was more confused. The demo took place the day that the Paris Agreement was adopted unanimously, amid the self-congratulations of world leaders. The demo supposedly rejected the Paris Agreement as insufficient, but the tense mood of the march (owing to the criminalisation of mass gatherings by the government following recent terror attacks in Paris) quickly morphed into celebration as the mobilisation was permitted. The crowds moved peacefully through Paris' streets, ending under the Eiffel Tower with a party-like atmosphere: music, song and dance. Though empowering for first-time activists to defy state repression of protest, this was jarring for those of us focussed on the collective suicide pact just signed by world leaders.

---

1  John Jordan, 'On "D12", we will draw our red lines in Paris', *Red Pepper*, https://redpepper.org.uk/on-d12-we-will-draw-our-red-lines-in-paris/ (last accessed 27 December 2020).

This is just one moment in the relatively brief history of climate activism, but for me it sums up where we are going wrong. It was a mistake to focus mass-mobilising capacities on the COP process which has such limited scope to deliver justice and over which civil society has such little serious influence (a symptom of the systems which produce climate injustice). The Indigenous Environmental Network (IEN) broke away from the coalition organising the main demonstration, instead performing a ceremony outside the Notre Dame cathedral earlier in the day, celebrating anticolonial resistance and protesting the Paris Agreement's neglect of Indigenous rights.[2] While the main demonstration was unexpectedly facilitated by the state, the police dispersed the IEN's gathering, meaning the ceremony had to be concluded on a nearby bridge. This should be unsurprising given the IEN's explicit rejection of the UNFCCC as a colonial, capitalist mechanism with no regard for Indigenous lives. Compare this to the coalition of mainstream NGOs registering their dissatisfaction at the Paris Agreement while levelling no fundamental critique of the process' inbuilt injustices. In Paris, the mainstream climate movement lacked coherent or proportionate demands for climate justice. The demonstration performatively gestured towards the movement's own (ultimately unfulfilled) strategic re-orientation while 'bearing witness' to injustices it had built no power to resist, let alone demand an alternative to.

## Does direct action get the goods?

When delivering non-violent direct-action training workshops to student campaigners, I stick ten photos of different actions

---

2   Wainwright & Mann, *Climate Leviathan*, Chapter 7, Section 1.

on the wall around the room. I ask people to stand by whichever they like the most and explain if they think it is direct action. The gimmick is that they are all direct action: blockading a fracking site, a tenants' union forming a human chain to stop an eviction, the Black Panther Party providing free breakfast for school children, a mass demonstration, Black Lives Matter pro-testors pulling down a statue of a slaver. I define direct action expansively as: 'taking action to address a social problem, rather than waiting for government representatives to do it themselves.'

In her book, *Direct Action*, L.A. Kauffman describes how pio-neering Earth First! and Greenpeace direct actions relied on a small number of people taking risky action. They stopped tree-felling 'by driving metal spikes into trees slated for logging' and prevented the slaughter of whales and baby seals as activists 'travelled to remote hunting grounds and placed their bodies between the killers and their prey.[3] Greenpeace's early actions were successful by combining high-risk disruption with a mass communications strategy to expose the public to striking images through the mainstream press and direct mails. As the public reacted with horror, legislative change restricting whaling and sealing followed. These were issues that could be feasibly resolved within capitalism. However, many of today's grassroots climate networks have inherited this specific model of direct action from 1970s and '80s environmental groups, seeking to apply it to a crisis which calls for a transformation of the whole economic system.

Between 2016 and 2018, I organised as part of the UK direct action network Reclaim the Power. I joined the 2016 'End Coal Camp' which culminated in a mass invasion of a large coal mine close to Merthyr-Tydfil in South Wales. I later took direct actions

---

3   L.A. Kauffman, *Direct Action: Protest and the Reinvention of American Radicalism* (London & New York: Verso Books, 2017), pp. 105–6.

against the fracking industry. Targeting Cuadrilla's fracking site at Preston New Road in Lancashire, and companies in the supply chain, the anti-fracking movement's strategy reflected early-Greenpeace. Small affinity groups of between 10 and 20 people planned secretive disruptive actions to shut down the fracking site (or a supplier) for around a day. Using a striking image of the action as a hook, we'd maximise press coverage while viral social media content replaced direct mails. Actions usually involved blockading the entrance to the site with activists locking themselves to each other through fortified tubes or another structure. The idea was that with enough of these disruptions, day after day, the site would not be sufficiently profitable to continue. And with an effective communications strategy, public opinion should swing against the industry.

In July 2017, I arrived with a small group to the campsite on a Sunday evening. We spent Monday arranging an action. The plan was for two people to cycle bike trailers towards the site from opposite ends of the main road and stop in front. We'd grab our fortified arm tubes from the trailer before jumping into the road (hopefully avoiding security and oncoming traffic). By 8am on Tuesday morning, I was lying face up in the middle of the major A-road connecting Preston and Blackpool. Only a precariously perched golf umbrella kept the torrential rain from my face. By lunchtime, I was extracted from my lock-on device by police, arrested (sodden and shivering) and taken to Blackpool police station. The fracking site was back up and running for the afternoon. By December, I received a conditional discharge for Obstruction of the Highway after two visits to court.

The campaign against Cuadrilla's fracking site relied on a churn of activists to get arrested, money to pay for increasingly extravagant blockades, and importing activists from around the country to set-up camp for days or weeks at a time. Amid challenging geological and economic conditions for kick-start-ing the fracking industry, the campaign of disruption certainly contributed to ensuring Cuadrilla never drilled commercially. In November 2019, the government indefinitely suspended fracking in England, though it was not permanently banned.[4] If a more financially resilient company invested in dozens of sites at a time, rather than just the one, and if the government permitted it, there's a good chance they could overwhelm the movement's capacities for direct action. With its focus on a limited set of exclusive and resource-intensive tactics, the anti-fracking movement did not scale or build the powerful mass movement required to reliably resist fossil fuel expansion in the UK.

Reclaim the Power emerged from Climate Camp: a series of camps between 2006 and 2010 attended by hundreds of people, targeting sites of key fossil fuel infrastructure from coal mines to Heathrow airport. This was a brand of direct action that mobilised through the camps, which either had a large associ-ated action (like invading a coal mine) or acted as a base for the smaller affinity-group actions. Like Occupy in 2011,[5] the camps were organised by anarchist principles and strived for horizontal organisation which rejected hierarchical structures of authority and leadership in favour of a system of direct democracy where every member had an equal say. Consensus decision-making processes aimed for everybody to be satisfied with a decision that was taken. At Occupy this took place at regular general assemblies where any member could bring a proposal, or block somebody else's. Climate Camp and then Reclaim the Power defined grassroots climate direct action in the UK for over a

---

4  'Fracking halted after government pulls support', *BBC News*, https://bbc.co.uk/news/business-50267454 (last accessed 27 December 2020).

5  Occupy was a global movement of occupations of public squares to protest in-equality, originating in New York City.

decade and inspired similar camps around Europe. The now regular Ende Gelände camps mobilise thousands of activists from across Europe against German coal mines. Also organised horizontally, the camps last a few days at a time, during which activists disrupt the coal mines using blockades to hold their position. The size and regularity of the camps and actions have successfully highlighted coal extraction across Europe and inspired similar actions in neighbouring Poland and Czechia. In 2016, the blockade of the Welzow-Süd coal mine and Schwarze Pumpe power station was reported as being part of 'the largest ever global civil disobedience against fossil fuels' as activists in countries around the world took similar actions.[6]

Direct actions like the End Coal Camp and Ende Gelände are logistically impressive operations and I've met many activists who entered the movement through mass camps. They are, however, also typical of the activist trend that Nick Srnicek and Alex Williams criticise as folk politics in their book *Inventing the Future*. Folk politics is a fetishism of the organisational approach of horizontalism and direct democracy, and the strategic focus on direct action applied to all contexts. They argue: 'direct action often remains insufficient to secure long-standing change, and in isolation, is typically only a temporary impediment to the powers of state and capital.'[7] This is true of this model of direct-action camps. Major energy companies operating coal mines can integrate these disruptions into their business models, comfortably absorbing the shocks of a few days of direct action. Movements that privilege immediate

6   Oliver Milman, '"Break Free" fossil fuel protests deemed "largest ever" global disobedience', *Guardian*, https://theguardian.com/environment/2016/may/16/break-free-protest-fossil-fuel (last accessed 04 May 2020).

7   Nick Srnicek & Alex Williams, *Inventing the Future* (London & New York: Verso Books, 2017), p. 28.

gratification through direct action and direct democracy (rather than the longer struggle of democratic politics and labour movement organising) do not scale to confront the capitalist system behind fossil fuel infrastructure or build power to win a coherent alternative.

Naomi Klein gives a name to the global wave of direct action resisting fossil fuel infrastructure: Blockadia. These resistances have targeted specific extraction projects through local organising combined with (inter)nationally coordinated mobilisation. As well as emissions, they have prioritised land rights, sovereignty and local livelihoods, as governments and corporations build infrastructure through Indigenous territory, farmland, water sources and communities. The Carmichael coal mine was first proposed in 2010 in Australia. Indian company Adani planned to build one of the largest mines in the world in the Galilee Basin in Central Queensland. The mine is to be built on the sacred traditional land of the Indigenous Wangan and Jagalingou people who have led the resistance against it. A combination of direct action and public campaigning has limited Adani's progress. Blockades have frustrated construction and campaigns targeting banks, insurers and suppliers, having caused dozens to withdraw their support for the mine. This organising has forced Adani to downsize the planned size of the project from AUS$16.5 billion to a self-funded AUS$2 billion.

In North America, Native Americans and First Nation Canadians have led opposition to the Keystone XL and Dakota Access pipelines (DAPL). The Standing Rock camp was the base for the resistance to Dakota Access in 2016 and became the largest mobilisation of Native Americans in over 100 years. This was archetypal Blockadia. The resistance attracted global media attention and solidarity as marches, direct action and spiritual resistance faced violent repression from police and private security. It took historic, protracted campaigns against both pipelines for President Obama to eventually delay them. In his first week in office in early 2017, President Trump used executive powers to restart the expansions. In July 2020, a judge in Washington DC suspended use of DAPL while the Supreme Court blocked construction of parts of the Keystone XL expansion. When Joe Biden took power in January 2021, he revoked the permit to Keystone XL by executive order on his first day in office.[8] The back and forth on these fossil fuel infrastructure projects highlights the potential of movements against them to drag centrist politicians in line. It also highlights the uncertainty that will always shroud these victories. The US may be another Republican President away from the pipelines receiving a new lease of life.

When Blockadia has been most successful, it has frustrated fossil capital's relentless drive for extraction in some of the most socially and ecologically egregious cases. But its victories have generally been relative and precarious, downscaling or delaying construction, or shunting decisions about major infrastructure into the legal system and out of democratic politics. And for each fossil fuel project Blockadia faces down, there are so many more forced through despite activists' best efforts, or subject to no organised opposition at all. Activists have grown confident in taking on fossil capital but have also defaulted into a defensive strategy of selective resistance. It can feel like climate activists are constantly putting out fires, while governments and corporations have a never-ending supply of kerosene and matches to set more alight.

---

8   Ben Lefebvre and Lauren Gardner, 'Biden kills Keystone XL permit, again', *Politico*, https://politico.com/news/2021/01/20/joe-biden-kills-keystone-xl-pipeline-permit-460555 (last accessed 7 February 2021).

## Climate industry

Greenpeace, Friends of the Earth, Sierra Club. These are likely the first environmental or climate change organisations that spring to mind. Greenpeace has its roots in disruptive direct action, protecting animals from slaughter and opposing nuclear weapons testing. Sierra Club was founded to preserve nature through national parks. Friends of the Earth was founded in 1969 in the US after splitting from Sierra Club due to Sierra Club's support for nuclear energy. All have evolved to become sizeable corporate NGOs with chapters around the world. They are, however, just three in an ocean of professional organisations of various shapes and sizes around the world working against climate change. I call this the environmental NGO industry. The rise of these organisations has paralleled the ascendancy of the financial sector under neoliberalism. Like finance, the NGOs don't produce goods or services of obvious economic value. And unlike direct-action networks' fetishism of alternative organisational forms, NGOs default to the organisational structure of a traditional corporation. They are hierarchical with a CEO or executive director at the top and a bloated stratum of middle managers. They have marketing departments to promote the brand, HR departments to oversee frequent restructures and no democratic process to empower staff or members. Professionalising the climate movement is supposed to add capacity and expertise. This often works when NGOs fill in the gaps left by more autonomous organising, including research, lobbying government and supporting legal action. This model fails, however, when NGOs suck up resources, dampen political ambition, and demobilise talented activists by subordinating them to organisational bureaucracy.

Just as neoliberalism depoliticises economics, the NGO industry can uphold the status quo by divorcing climate action from politics. The ideology prevalent across the environmental NGO industry is a mixture of anarchism, borrowed from grassroots direct-action networks, and the centrist liberalism of the corporate third sector. The anarchist influence provides a radical edge while feeding a reluctance to pursue state power or engage with electoral party politics. The liberal influence provides superficially progressive stances, for example on racism, sexism, inequality and global injustice, while limiting its political horizons to operate within capitalism. Though capitalism and colonialism may occasionally be critiqued, NGOs never go so far as to demand democratic public ownership or other socialist reforms (I further explore these in Chapters 5 and 7). In their liberalism, class is not an economic relation that structures the ownership of capital in the economy, and thus who profits from climate change and who bears the brunt of its injustices. Class, if mentioned at all, is added to a long list of identities from which someone may be privileged or oppressed.

The political limitations of NGOs are often put down to a hostile regulatory environment. In the UK, the Lobbying Act 2014 was introduced to restrict NGOs' lobbying in the run up to national elections. In reality, the restrictions are always in place because of the frequency of elections and length of the regulated lead-in time. Restrictions are even tighter in the US. Though legislation is often blamed for NGOs' political impotence, legal constraints and the industry's ideology are co-productive. In 2015, both Greenpeace and Friends of the Earth took 'civil disobedience' against the Lobbying Act by refusing to register as campaigners per Electoral Commission guidelines. Both received sizeable fines.[9] They were happy to pull a performative stunt (akin to a donation to the Electoral Commission) but were nowhere to be seen when the stakes were high. By the 2019 general election, the Labour Party's proposed Green New Deal (discussed in Chapter 5) encapsulated a more ambitious set of climate and environmental policies than mainstream NGOs themselves and the Green Party (according to Friends of the Earth rankings).[10] In this moment, NGOs like Greenpeace might have practised their tradition of civil disobedience and actively worked to elect a Labour government, despite repressive laws, or at least campaigned for a Green New deal independently. Instead, they embraced their apolitical comfort zone. In fact, a suite of NGOs including Greenpeace, Friends of the Earth and World Wide Fund for Nature (WWF) publicly supported a net-zero emissions target of 2045 – just five years sooner than the Conservative government's target – while Labour proposed 2030.

A more fundamental cause of NGO ideology (including their reluctance to seriously intervene in even the starkest general elections) is the industry's funding model. Larger NGOs with household recognition are generally funded by individual donors or legacies. Naturally, donors are politically diverse, so NGOs don't want to alienate donors by taking radical positions. NGOs' legal status as charitable organisations, which is useful for raising this revenue, makes them vulnerable to hostile

9  Kirsty Weakley, 'Friends of the Earth and Greenpeace fined for breaches of "unworkable" Lobbying Act', *Civil Society News*, https://civilsociety.co.uk/news/friends-of-the-earth-and-greenpeace-fined-for-breaches-of-unworkable-lobbying-act.html (last accessed 27 December 2020).

10  'Election manifestos: Labour tops Friends of the Earth's climate and nature league table', https://friendsoftheearth.uk/general-election/election-manifestos-labour-tops-friends-earths-climate-and-nature-league-table (last accessed 20 July 2020).

government or regulators which could revoke it should they become too political. For smaller NGOs, a majority (sometimes 100%) of their funding can be grants from philanthropic trusts and foundations. This is essentially the small change of wealthy individuals or corporations. If smaller NGOs were funded with no strings attached, they would be free to innovate with radical strategies and demands and push larger organisations further. In my experience, there are normally enough anti-capitalist staff in these organisations who desire such ambition that this would be entirely feasible. Instead, smaller NGOs and their staff are hamstrung by a dependence on grants from philanthropic capital. Why would capital (however progressive the individual or organisation) continue to fund an NGO forcefully organising against the capitalist system in which capital operates (even if it's crucial for climate justice)? Unless the NGO industry can shift its funding model to break from its legal vulnerability to the state and financial dependency on capital, it will remain limited in what it can do.

For some NGOs, campaigns to shift individual consumption are more appealing than taking on capitalism. The idea of an individual 'carbon footprint' (originally propagated by BP to distract from the responsibility of fossil fuel companies) guilts individuals into changing their behaviour without acknowledging that most peoples' behaviours are structured by the fundamentally unjust, high-carbon economic system they find themselves within. The 'greenest' lifestyle choices we're told to pursue involve spending more on consumer goods and taking time that most people don't have to spare. WWF's annual Earth Hour is a prime example. For just one hour of just one day, WWF encourages people to turn off their lights to conserve energy. This is often supplemented by the operators of major infrastructure, like large office buildings or monuments, turning off their lights to get in on the PR stunt and contribute to an impressive aerial video of lights in a major city switching off together. But climate justice isn't ordinary people sitting in the dark to conserve energy. It's abolishing fuel poverty by providing clean energy to everyone for free. As I strolled through the Westgate shopping centre in Oxford just before Earth Hour in March 2019, I saw a series of WWF adverts on digital billboards. They were displaying photos of iconic animals (tigers, penguins and walruses) next to text: 'We are the first generation to know we're destroying the world. But we can change this. Why don't you pledge to turn washing to 30°C, or change the way you eat?' This messaging distributes blame to our generation as a whole, rather than the capitalists and governments managing the economy. It limits the scope of possible solutions to the least impactful individual actions rather than structural transformations. NGOs like WWF have previously taken money from fossil fuel companies and spent the 1990s supporting neoliberal initiatives like the North American Free Trade Agreement (NAFTA). Today, they continue to uphold capitalism and the interests of fossil capital by deflecting blame away from them and lowering our collective ambitions.

## Another NGO is possible?

In July 2012, Bill McKibben wrote an article for Rolling Stone magazine titled, 'Global Warming's Terrifying New Math'. McKibben lays out the headline numbers of climate change: emissions, temperature rises, carbon dioxide particles in the atmosphere. He goes on to make the opening pitch for the fossil fuel divestment movement which went on to inject life into a stagnant climate movement. Telling the story of how divestment was used successfully as a lever of international pressure against the apartheid regime in South Africa, McKibben makes the case for universities, pension funds and investors to 'dump stock from companies that are destroying the planet' to weaken the fossil fuel industry's political standing.[11] The logic is that if fossil fuel companies like Shell or ExxonMobil are exposed for their role in climate injustice, they might be forced to stop lobbying against climate action, stop funding politicians or even transition to become clean energy companies.

350.org was founded in 2007 by Bill McKibben and a group of his students from Middlebury College. This was a new climate NGO with its name derived from the amount of carbon dioxide that can safely be in the atmosphere: 350 parts per million. Whereas other NGOs have evolved from their origins as environmental organisations, sometimes even happy to accept donations from oil executives, 350.org was founded in the age of climate crisis with climate justice a foundational goal. 350.org existed for five years prior to 2012, organising global marches, days of action, and campaigns for presidential candidates to improve their climate policies. But it was divestment that catalysed its rise. The Fossil Free campaign bucked the apolitical trend of the NGO industry by embracing an explicitly anti-corporate climate politics with an open hostility to the fossil fuel industry. Divestment has impacted the climate movement itself, forging an antagonism between the movement and fossil capital. That antagonism has become common in certain sectors like faith organisations, higher education, and philanthropy where the campaign is most successful. And it has penetrated national politics in the US as high-profile candidates succumb to grass-

11   Bill McKibben, 'Global Warming's Terrifying New Math', *Rolling Stone*, https://rollingstone.com/politics/politics-news/global-warmings-terrifying-new-math-188550/ (last accessed 6 May 2020).

roots campaigners' pressure to refuse donations from fossil fuel companies.

I spoke to Rob Abrams who led divestment campaigns at Swansea University and SOAS. I asked him whether he thought that the divestment movement was successful:

> The divestment movement has definitely achieved immense success, reaching milestones that few would have thought possible a few years ago. The trickier question is whether it achieved what it set out to. I found that it often wasn't clear whether we wanted to stigmatise the industry or bankrupt it altogether. I think individuals like Bill McKibben understood the way that coordinated moral outrage could, as he put it, 'give rise to a real movement'. In that sense, we can see that one of divestment's principle successes lay in re-orientating many environmentalists towards a socio-economic analysis of climate change and direct conflict with fossil capital.

What were the main divisions within the Fossil Free movement?

> The movement was ultimately an uneasy alliance between two main camps. The first was those wanting radical, structural change in our society and our economy. The second was those interested in reforming fossil fuel companies and banks to bring about a 'green capitalism'. The radical camp was motivated by the opportunity to undermine fossil capital and grow the power of the movement. The reformist camp positioned itself as the saviour of institutions that could reclaim their moral standing by acting as 'responsible investors'. Any unity that briefly existed between these camps couldn't be maintained in the long-term. Radical organisers were drawn to other campaigns which offered a more coherent path

towards a just and swift end to the fossil fuel industry, like a transformative Green New Deal.

Is there anything we can learn from the divestment model?

The main lesson that the divestment model can offer is simplicity. Fossil fuel divestment mobilised so many disparate campaigns and secured commitments representing trillions of dollars because at the heart of its theory of change was one simple equation: 'the fossil fuel companies want to burn X amount, but we can only burn Y amount if we want to avert a worst-case climate change scenario'. The course of action that then followed was relatively straight-forward and could be easily replicated to fit a number of different contexts. A visible 'bad guy' helped divestment groups to polarise their communities and provoke their targets to pick a side.

After divestment sowed the seeds of antagonism with fossil capital, the natural next steps would be to land material blows. Attempts have been made by campaigners to cut the supply of finance from banks for fossil fuels, block insurance for new infrastructure, end subsidies from government, marginalise the industry's grip on politics, and spur investment in alternative energy systems. Until now, though, the many trillions of dollars divested from fossil fuels have not been translated into serious attempts to dismantle the industry. Major banks like Barclays and HSBC, subject to grassroots and NGO campaigning, have tweaked their energy policies to exclude finance for coal and tar sands projects, generally where they were already least implicated. They still finance the companies directly and make no indications of abandoning other oil and gas.[12]

---

12   Banking on Climate Change: Fossil Fuel Finance Report 2020, https://www. ran.org/bankingonclimatechange2020/, accessed 27 December 2020.

350.org is the exception that proves the rules of the environmental NGO industry. In leading fossil fuel divestment globally, the organisation has pushed the limits of NGO ideology. 350.org has called to dismantle the fossil fuel industry (a core fraction of global capital) and even amplify the abolitionist demand to defund the police, in solidarity with Black Lives Matter (the police being a key state institution in defending private property and capital). At the same time, it has not drawn the links between those two demands and the need for a coherent program to supplant capitalism with socialism (or anything in its image). Like other NGOs, the corporate structure of 350. org has allowed it to scale as a truly international organisation. In doing so, it has created a global movement, but not a mass movement. 350.org organises with a model of distributed organising without a social base. To win a local divestment campaign, a chapter comes together around the cause on a university campus or targeting a city council, but is not bound together by anything other than the temporary campaign. When they win, the group either withers away or shifts focus to some other technical or hyper-local issue. Without a compelling story to tell of how to win climate justice beyond divestment, the movement's energy dissipates.

The coalition of NGOs and direct-action groups which organised that anti-climactic march through Paris in 2015 pledged to take the climate movement to the next level. Years later, some of those organisations continue to contribute usefully to the broad movement, but none have provided the leadership and direction needed. In the final years of the 2010s, a new generation of climate organising brought fresh energy to the movement. Although it manifested in different approaches, the move to a new type of climate politics was underway. These new groups brought urgency and disruptive tactics, new bases of activists, and radical demands. Would this be a breakthrough moment in the fight for climate justice?

# Chapter 4

# The next generation

On the 31 October 2018, over 1,000 people gathered at Parliament Square and 'declared [themselves] to be in open rebellion against the UK government'.[1] This was the beginning of Extinction Rebellion (XR). Organisers worked to set up hundreds of local XR groups in the smallest towns to the biggest cities across the UK and around the world. In November 2018, XR took the world by storm with their first International Rebellion. London was the focal point as activists blocked roads to shut down Oxford Circus, Marble Arch, Waterloo Bridge, Piccadilly Circus and Parliament Square. Over 1,000 people were arrested over ten days of disruption. XR's demands were simple. First, the UK government should 'tell the truth' by declaring a 'climate and ecological emergency'. Second, the government should act accordingly and decarbonise the UK economy to reach net-zero emissions by 2025. Third, the government should form, and be led by, a citizens' assembly of randomly selected members of the public to deliberate on a response to the emergency. XR dominated the mainstream news agenda in the UK for a sustained period. This was new. These were the most ambitious demands for total decarbonisation seriously levelled at the government. It was the first time generalised social and economic

---

1   Sam Knights, 'Introduction: The Story So Far', in *This is Not a Drill* (Milton Keynes: Penguin Random House UK, 2019), p. 10.

disruption was used by climate activists over a sustained period with such high participation.

In the summer of 2018, attitudes towards the climate crisis began to shift as the severity of the crisis hit home. Many people already believed that we needed to do something about climate change, but if this manifested in action it was usually limited to individual behaviours such as recycling, rather than engaging in broader activities such as collective campaigning. Wildfires spanning the globe that summer were not new but penetrated the public consciousness unlike before. Fires raged most famously in the Arctic Circle, California and Athens, plus across England and Wales from June into September. Images of climate impacts in the Global North became unavoidable. In August, leading scientists warned of a future 'hothouse earth' scenario. This meant that we are reaching planetary tipping points faster than expected, unleashing new sources of emissions, and accelerating climate change so that human mitigation efforts would soon become futile. Into October, the IPCC published a report with the *Guardian*'s headline summarising the key message: 'We have 12 years to limit climate change catastrophe, warns UN'.[2] The BBC and *Guardian* revised their editorial priorities, giving greater prominence to climate change stories. The emergence of XR and the youth strikes coincided perfectly with this. They captured a popular mood of newfound urgency and channelled it into action.

On 20 August 2018, Greta Thunberg went on strike from school to protest inaction on climate change. She began by sitting outside the Swedish Parliament every day, and then weekly on Fridays. Thunberg's individual action birthed a movement

---

2  Jonathan Watts, 'We have 12 years to limit climate change catastrophe, warns UN', *Guardian*, https://theguardian.com/environment/2018/oct/08/global-warming-must-not-exceed-15c-warns-landmark-un-report (last accessed 17 March 2020).

known by different names: Fridays for Future, youth strikes for climate, or 'the school strikes'. After a smattering of similarly small actions, Australia took the strikes to the next level with 15,000 school children mobilised in November 2018. The Australian Youth Climate Coalition (AYCC) devoted significant resources to train kids and develop the movement to include a clear set of political demands, including opposing the proposed Adani Carmichael coal mine and the federal government's promotion of fossil fuels. In each of Germany, Switzerland, and Belgium, up to 70,000 kids protested in January 2019. 15,000 went on strike in the UK on 15 February 2019. The first global strike was 15 March 2019, with the largest turnouts including 195,000 across France, 200,000 across Italy, 300,000 across Germany. By the September 2019 global strikes, 7.9 million people were mobilised as adults took part in solidarity. Could this be the moment that a global movement became a mass movement?

It's significant that these were strikes and not simply bog-standard marches. Borrowing from the tactical repertoire of trade unions, children around the world displayed an intuitive understanding that they have power in collective action. In the UK, youth strikes contained an instinctive political radicalism, distinguishing them from the wider climate movement. Charlotte England and Clare Hymer reported for Novara Media that chants heard at the first youth strike in Brighton and London included: 'fuck Theresa May', 'fuck the Tories!' and 'Tories, Tories, Tories, out, out, out!'[3] Those initial strikes felt different to previous climate mobilisations because they mediated an expression of

---

3 Charlotte England & Clare Hymer, 'Fighting for the Future: How the Instagram Generation Are Renewing the UK Climate Movement', *Novara Media*, https://novara media.com/2019/04/12/fighting-for-the-future-how-the-instagram-generation-are-renewing-the-uk-climate-movement/ (last accessed 17 March 2020).

pent-up anger directed towards clear political enemies: the government, the Prime Minister, the Conservative Party. For the first time, climate injustice was the focal issue around which a generation expressed its broader disdain for the ruling class.

## Beyond politics

Greta Thunberg's position as founder of the youth strikes movement, combined with the power and clarity of her messages, have helped her to cut through as the definitive voice of a generation. Her approach to the climate crisis is similar to the founding ideas of XR. In *No One is Too Small to Make a Difference*, Thunberg writes: 'This is not a political text. Our school strike has nothing to do with party politics.'[4] Instead of committing to a specific ideological response to the climate crisis, both Thunberg and XR appeal heavily to the authority of science and 'the truth'. At the Stockholm Climate March in September 2018, Thunberg begins by referencing scientists' warning that we have limited time to reverse growth in emissions. 'If people knew this they wouldn't need to ask me why I'm "so passionate about climate change".' The underlying logic here is that knowledge leads to action. This was foundational to XR's original mobilisation model, in which 'the talk' (given across the country to recruit activists) began with a lecture on the stark science of the climate emergency.[5] XR's deference to 'the science' and calls to 'tell the truth' are a core part of its founders' project to remake climate activism. The headline of XR's third demand is 'beyond politics'. More than just a demand, for XR co-founder Roger Hallam, it is a foundational value for a new climate movement

---

4   Greta Thunberg, *No One Is Too Small to Make a Difference* (Milton Keynes: Penguin Random House UK, 2019), p. 2.

5   Knights, *This is Not a Drill*, p. 10.

and a reaction against what he calls 'the reformist political class' (apparently all political parties, NGOs, and most campaigners) which has failed to stop climate change.

He argues that this failure is down to the domination of environmentalism by 'the left' and political ideology: 'The bit-by-bit reformist framing of change is both immoral and ineffective as it puts political ideology before scientific facts.'[6] Hallam believes that any commitment to political ideology obscures the urgency of following the science and its logical endpoint of overthrowing the government. This hostility to ideology informed XR's initial refusal to adopt a detailed programme of demands beyond three headlines. Calls to 'tell the truth' and reach net-zero emissions by 2025 are deliberately vague, offering no recommendations for how to go about them. The demand for a citizen's assembly, is the most practical, but acts as a get-out clause for the imprecision of the other demands. It abdicated responsibility for justifying their demands as workable and campaigning for specific measures. Instead, it defers that work to an undefined future group of randomly selected citizens. This might make sense if climate change was simply a scientific problem. But climate change is a political crisis born out of capitalism and requires a proportionately ideological response.

'Beyond politics' was at the same time crucial to XR's explosive rise and a prime strategic limitation. XR was successful in mobilising an unprecedented number of people to channel their climate anxieties into civil disobedience by presenting a clear, stark problem alongside a simple but impactful action to take (block a road and get arrested). By avoiding contentious issues in messaging and demands, people of all political affiliations and none felt able to take part without compromising their values. It

---

6   Roger Hallam, *Common Sense for the 21st Century* (Carmarthenshire: Common Sense for the 21st Century, 2019), p. 55.

created a space for many people who had never been 'activists' before to make some contribution to the existential issue of the day. The aversion to politics was limiting, though, when it came to consolidating victories or making material progress. In the UK, over 300 (that's almost three quarters) of local authorities have declared a climate emergency. Yet the response to climate change in local government has not significantly shifted because XR lack the political ideas to translate moments of 'telling the truth' into action. Citizens' assemblies have been hosted by UK Parliament and local authorities like Oxford City Council. In both cases, the recommendations were mostly a list of ideas we already had floating around but without a cohering plan or binding responsibility to implement them. In these examples of XR's headlines being half-met by national or local government, they have simply been co-opted to give the impression of action by authorities with little intention of taking any serious material actions. XR created a much-needed space in the mainstream to discuss the climate crisis and its injustices. Undoubtedly, 'beyond politics' allowed it to achieve what it did where other climate organisations have failed to capture the public debate and mobilisation on such a scale. However, founding ideas like 'beyond politics' soon became points of contention within the young organisation and the wider movement.

## XR in action

In its early stages, XR converted interest in its alarming account of the climate emergency into action with a clear story of what needs to be done. The founding strategy was heavily informed by Roger Hallam's ideas. In his book, *Common Sense for the 21st Century*, it's clear that he is at heart a classical liberal. For Hallam, capitalism as a system is not the fundamental problem, but how the state's role in managing it has been distorted. 'The problem resides with the state and its capture by the corporate business class.'[7] If the public can reassert its democratic control of the state, then it can return to its proper functions. XR's founding proposition was to overthrow the government and institute a citizens' assembly in its place. With these ambitious goals, what was the plan? Unlike the failed tactics of traditional climate campaigns (for example, A-to-B marches and petitions), XR promoted a strategy of non-violent civil disobedience, economic disruption and mass arrests. The pitch to supporters was simple. If they felt compelled to take action about the climate emergency, they could join a blockade of a road, bridge or key infrastructure and get themselves arrested.

XR's mobilisation strategy was based on social movement research suggesting that with the support of about 3.5% of the population, through targeted disruption in the capital city, a rebellion can succeed in overthrowing the government.[8] Activists getting arrested en masse for disruptive actions would supposedly overwhelm the state and win popular sympathy for the cause. This was justified by pointing to examples of revolutions overthrowing dictators throughout history, from the British occupation of India to the Arab Spring. Farhana Yamin, a climate lawyer who had been a lead author on several IPCC reports and joined Extinction Rebellion, wrote: 'We need everyone to undertake mass civil disobedience to create a new political reality the whole world over.'[9] The problem is

---

7 Hallam, *Common Sense*, p. 10.

8 'The "3.5% rule": How a small minority can change the world', *BBC Future*, https://bbc.com/future/article/20190513-it-only-takes-35-of-people-to-change-the-world (last accessed 31 December 2020).

9 Farhana Yamin, 'Tell the truth. Die, survive or thrive?', in *This is Not a Drill: The Extinction Rebellion handbook* (Milton Keynes: Penguin Random House UK, 2019), p. 22.

that the examples cited may have led to new political regimes, but not economic transformation. Climate justice will require building a new economic system, not just changing who governs capitalism.

At the November 2018 rebellion, thousands shut down the five major bridges over the River Thames. A *Guardian* video records Hallam suggesting to police that they arrest protestors faster by requisitioning buses to transport them away (which they do). Hallam tells the cop: 'Because we don't really want to block the roads. We just want to get a load of people arrested and then we can say to the politicians, you know, a thousand people are happy to lose their liberty because we want some change.'[10] The April 2019 rebellion escalated attempts at economic disruption and mass arrests. In London, the fortnight's most iconic images were a bright pink boat blockading Marble Arch and the ethereal transformation of Waterloo Bridge into an urban garden, prefiguring the possibility of a post-car metropolitan ecology. After thousands of arrests and several days of serious disruption to the UK's capital city, the government was not overthrown. The rebellions' main effect was to dominate the news agenda, culminating in the UK Parliament declaring a climate emergency.

XR's tactical repertoire expanded as the movement grew. Activists began to autonomously organise more symbolic actions and campaigns which tended to make a similar media impact as the more disruptive actions. Earlier in April 2019, eleven rebels held a naked demonstration inside the House of Commons

---

10   Bruno Rinvolucri, Irene Baqué, Christopher Cherry, Adam Sich & Katie Lamborn, '"We can't get arrested quick enough": Life inside Extinction Rebellion – video', *Guardian*, https://theguardian.com/world/video/2018/nov/22/we-cant-get-arrested-quick-enough-life-inside-extinction-rebellion-video (last accessed 19 June 2020).

during a Brexit debate, gluing themselves to the glass window of the Gallery. The protest received widespread coverage as the shocked reactions of MPs, including the wide eyes of former Labour Party leader Ed Miliband, adorned social media for the next 24 hours. A group organising for global justice emerged within XR. This was organised by activists who recognised the potential of XR but also recognised gaps in its political outlook. At the same time, actions became more targeted with increased focus on the fossil fuel companies and financial institutions most responsible for climate injustices. Farhana Yamin was arrested for gluing herself to Shell's headquarters, Sam Knights was arrested protesting an oil and gas conference, and protests at Barclays bank became commonplace. In May 2019, UK Parliament declared a climate emergency on the back of XR's November and April rebellions. Knights argues: 'It remains the largest victory that Extinction Rebellion can legitimately claim.' [11] He and Yamin had coordinated XR's political negotiation strategy which led to meetings with Conservative minister Michael Gove and Labour Shadow Chancellor John McDonnell on the day of the vote. This was an achievement of a headline demand, but it required a level of engagement in the formal political process which some felt betrayed XR's founding values.[12] Some of these tactical developments took place within XR's model of civil disobedience provoking arrest, but they also indicated a shift away from the 'beyond politics' principle and strategy of generalised economic disruption that had originally defined XR as unique.

---

11   Sam Knights, 'Extinction Rebellion: We Need To Talk About The Future', *Medium*, https://sam-j-knights.medium.com/extinction-rebellion-we-need-to-talk-about-the-future-95459aa4d4e0, (last accessed 25 March 2021).
12   Ibid.

One year after its launch, the aim of overthrowing the government had not been realised. While most members were content with XR's new orientation, a small minority were committed to the founding strategy of escalating general economic disruption. In June and July 2019, a group planned to use drones to shut down Heathrow airport during the summer holidays. It was eventually called off after fierce internal debate with some members threatening to leave XR if it went ahead.[13] During the October 2019 Rebellion, a small group jumped on top of a morning rush-hour London Underground train at Canning Town station. The world saw viral video clips of angry commuters dragging XR activists off the top of the train, furious at the disruption to their commute. The action received widespread condemnation for disrupting the ability of low-wage workers to get to work and targeting low-carbon public transport while protesting climate change. Sam Knights reflected:

The infamous train action is, I think, a good case study in the limits of decentralised activism. It was planned by a tiny faction of the movement, opposed by the overwhelming majority of our activists, and yet it was still allowed to happen under the banner of Extinction Rebellion. It soaked up all the media coverage that day, distracting from more thoughtful actions and altering the public perception of our protests. Yet it was only about ten or fifteen people who were ever convinced it would work.[14]

---

13  Sanjana Varghese, 'Extinction Rebellion's Heathrow drone protest is tearing it in two', *Wired*, https://wired.co.uk/article/extinction-rebellion-heathrow-air-port-drones (last accessed 31 December 2020).
14  Knights, 'Extinction Rebellion'.

This was the final straw. When XR launched its 2020 strategy, it was clear that the already fringe faction responsible for Canning Town had been marginalised completely. It declared: 'In 2019 we demanded change. In 2020 we begin building the alternative.' The plan was for new focuses on prefiguring demands by 'piloting new participatory systems in democracy, media and economics' and supporting targeted mobilisation including against aviation expansion, fracking and coal. COVID-19 meant that they couldn't implement the strategy in full, but what XR did manage amid the pandemic confirmed a shift towards more targeted and political campaigning. Direct action mostly came through 'HS2 Rebellion', targeting the construction of the HS2 high-speed rail project. 'Money Rebellion' consolidated the focus on finance, including the targeting of Barclays by organising actions against its fossil-fuel financing. XR's political work developed further with a campaign for a Climate and Ecological Emergency Bill which was brought to Parliament as a Private Members' Bill (the focus of updating the UK's climate targets and implementing citizens' assemblies). Do these developments signal XR's growing maturity as it receives and responds to criticism by evolving? Or are they a worrying sign that XR has been worn down, cowering away from its niche? Has XR has become too similar to the traditional climate organisations and campaigns it was founded as an alternative to?

## Generation climate

In December 2019, Greta Thunberg told activists in Madrid that the school strikes had achieved nothing.[15] In one sense, she's right.

15  Fiona Harvey, 'Greta Thunberg says school strikes have achieved nothing', *Guardian*, https://theguardian.com/environment/2019/dec/06/greta-thunberg-says-school-strikes-have-achieved-nothing (last accessed 20 March 2020).

The youth strikes experienced an initial explosion, dragging climate change into the mainstream of media and politics. But the pandemic disrupted the movement's momentum, and, like XR, they cannot yet take credit for any practical action for climate justice. Speaking to youth strikers, it's apparent that they put a lot of pressure on themselves. Heavily critical of the failures of 'adults' to stop climate change, they shoulder the burden of protecting their own futures. Not yet instigating the profound economic transformations necessary for climate justice does not mean the strikes have 'failed'. They have laid the groundwork for a powerful political generation only just beginning its journey.

The intergenerational injustice of climate change is obvious when you look at the distribution of climate impacts over time. Many of capitalism's injustices are felt more immediately; the exploitation of workers by their boss is an everyday experience. So is the racist violence of police forces, living under colonial occupation, and food or fuel poverty. With climate change, fossil fuels burned tens or hundreds of years ago produce impacts in an entirely different place and time. As greenhouse gases concentrate in the atmosphere, climate impacts like flooding, extreme weather and drought become more frequent, severe and geographically spread. Those who profit from building and maintaining the fossil fuel economy may not even live to witness (let alone endure) the worst effects of their business model. Those who have been (and will be) born into this economy and this climate are guaranteed to live through years of devastating climate change. For many youth strikers, this is an all-consuming prospect which defines their generation. As Thunberg has lamented: 'When you think about "the future" today, you don't think beyond the year 2050. By then, in the best case, I will not have even lived half my life. What happens next?'[16]

---

16   Thunberg, *No One is Too Small*, p. 10.

For today's teenagers and young adults, generational politics means knowing that they will live through the traumatic effects of decisions taken by today and yesterday's ruling class. It means 'the future' ending halfway through your expected lifespan because nobody dares imagine what the world looks like 30 years from now. And it's not just the climate crisis that younger generations have to navigate. Keir Milburn writes in his book *Generation Left:* 'Millennials are likely to be the first genera-tion to earn less than the two generations who came before.'[17] The COVID-19 pandemic has further precipitated a crisis of unemployment which, like austerity, young people will feel the hardest. Though youth strikers (part of the tentatively named Generation Z, born 2001–2015) come after Millennials (born 1981–2000), they will experience the same economic condi-tions of low-wages, insecure work, high rent, low prospects of property ownership, expensive higher education, and dimin-ished access to collective social security and welfare.

Though the generational injustices of climate change are clear, and youth strikers are justified in drawing attention to them, generation cannot be the prime lens through which we interpret the climate crisis. We have seen already that the injus-tices of climate change are being endured right now across the world with class and race the primary determinants of who is hit by those impacts. In 2019, Cyclone Idai destroyed 90% of the city of Beira in Mozambique.[18] When flooding hit the south-east of England and Yorkshire in 2020, residents were left with their homes devastated, while no firms were willing to insure for the damage. There is of course a generational element to these

17   Keir Milburn, *Generation Left* (Cambridge: Polity Press, 2019), p. 8.
18   '"Beira city '90 percent destroyed" by Cyclone Idai, hundreds dead', *Al Jazeera*, https://aljazeera.com/news/2019/3/18/beira-city-90-percent-destroyed-by-cyclone-idai-hundreds-dead (last accessed 1 January 2021).

injustices as past emissions drive present impacts, but today's corporations are committing human rights violations against today's global poor and colonised peoples. The people of West Papua live under the violent occupation by the Indonesian state enabling their land to be exploited by western mining corporations like BHP today. People living in La Guarija in Colombia are displaced by private militias to clear space to expand the largest coal mine in Latin America right now. 'Older generations' are not political agents in and of themselves. 'The adults' are no more a coherent agent culpable for climate injustice as 'humans'. It is capital and the ruling class, via capitalism, that is to blame for inflicting the harms of climate change onto the poor and colonised within their own generation, as well as future generations.

Milburn concludes *Generation Left* on a conciliatory note for older generations: 'It's easy to rage at the Baby Boomers when we look at the regimes they've voted into power, but we must remember that this is simply what a defeated generation looks like.'[19] Today's climate crisis is one symptom of decade upon decade of capital asserting its dominance over labour and nature, defeating successive political generations in the process. Baby boomers in the UK may disproportionately own property (and vote accordingly), but the lack of home insurance for flooding victims shows how climate change makes property ownership precarious. As climate change intensifies, the heat of summer and cold of winter will be more deadly for pensioners. A parallel crisis of demographic ageing combined with neoliberalism's erosion of the welfare state heralds a near future of greater risk of destitution and loneliness for pensioners. A radical generational politics in the age of climate crisis acts in solidarity (not

---

19    Milburn, *Generation Left*, p. 124.

opposition) with older generations by directing our rage at the ruling class which would prefer us divided.

I asked Scarlett Westbrook, a school striker and climate justice activist, whether she thought the youth strikes movement has been successful.

> The school strike movement has achieved some incredible things including contributing to the UK government's decision to declare a national climate emergency, increasing public awareness of the climate crisis, and paving the way towards other dynamic youth climate initiatives. However, the lack of organisational structure, the absence of a shared end goal of climate justice via policy frameworks like the Green New Deal, entrenched institutional racism, and unsustainable protest tactics limited the movement's success. This movement was the catalyst to broader organising in other forms, despite not being the solution itself.

What were the biggest challenges the movement faced?

> The movement has faced huge challenges every step of the way. The most significant problems resulted from an absence of a clear end goal, a failed organisational structure which led to unaddressed systemic issues with race and class, exacerbated by Southern-centricity and a knowledge hierarchy. We didn't make enough efforts to ensure that our tactics and structures were sustainable. I think we were let down by supporting adults who seemed more focused on newspaper headlines than the continuity of our movement, neglecting internal problems and leading to these issues growing more prominent.

Where does this politicised generation go from here? How can it channel that initial momentum?

> As largely disenfranchised young people, we cannot exert our influence through traditional means. However, we've created a precedent with the school strikes of making our voices heard through direct action, and we can continue to do this in other forms. We now have a platform for youth activism and hindsight from dealing with past issues. Young people will continue to protest through organisations like Teach the Future to influence policy, and local strike groups will pressure the government via direct action. We can unite these groups with the vision of a global Green New Deal and fight together to win.

Milburn discusses the importance of events in the class formation of generations. There are 'passive events' which are done to us and generally foster conservatism, restricting a widespread sense of political possibility. Then there are 'active events' where 'participants experience something they have actively constructed with others.'[20] In the formation of Generation Left (the politicised Millennial generation), the 2008 financial crisis was the passive event followed by the 2011 global wave of protest (Arab Spring, Occupy, UK student uprising) as the active events. If Generation Left was formed by student uprisings and functioned as the base of Corbynism, are we now seeing the formation of its successor: Generation Climate? The summer of 2018 was a long passive event, instilling fear into the generation: wildfires, IPCC report, hothouse earth report. Their active event was the youth strikes movement, giving the gen-

---

20    Ibid., p. 58.

eration a sense of power and possibility. Generation Climate remains in the process of forming. The youth strikes as a specific formation were no silver bullet. As Westbrook highlights, the youth strikes had to deal with their fair share of issues including organisational structure, internal racism and relationships with NGOs. The student uprisings resisting the tuition fee hikes faced similar challenges and that wave of action did not end the marketisation of higher education in the UK. However, Generation Left matured to become the base of democratic socialist projects like Corbynism. The youth strikes could similarly catalyse Generation Climate to mature to adopt a more propositional set of demands, a more enduring organisational form and a healthier internal culture. Milburn writes that 'political generations are intimately entwined with the dynamics of class struggle.'[21] Youth strikers are already somewhat familiar with the traditions of the labour movement having borrowed the strike from its tactical repertoire, normally used by workers against capital. Do the school strikes foreshadow a lifetime of ecological-class struggle for Generation Climate? Climate injustice is a symptom of class warfare waged by the rich against the global poor and working-class, the colonised, young and future generations. If Generation Climate can ally with Generation Left (there is already some overlap in personnel), then together we can co-develop an alternative and a strategy to strike back against the capitalists intent on stealing our future while exploiting us in the present.

---

21    Ibid., p. 106.

# Chapter 5

# Green New Deal – a blueprint

When Alexandria Ocasio-Cortez burst onto the scene in US politics, she brought the Green New Deal front and centre. In June 2018, she defeated incumbent Joe Crowley to run as the Democrats' nominee in New York's fourteenth congressional district. In November 2018, she comfortably beat her Republican opponent. Before she was sworn in, Ocasio-Cortez joined activists from the Sunrise Movement (a youth-led organisation campaigning 'to stop climate change and create millions of good-paying jobs in the process') as they occupied the office of Nancy Pelosi, the Democrats' leader in the House of Representatives, calling for a Green New Deal. Shortly afterwards, Ocasio-Cortez and Senator Ed Markey published their resolution for a Green New Deal: a 10-year mobilization of investment and regulation to create green jobs, decarbonise the economy, and address social and economic injustices.

They were not the first to call for a Green New Deal though. The term was coined by New York Times journalist Thomas L. Friedman, and a group of British environmentalists and economists first put meat on its bones with a report in July 2008 calling for 'joined-up policies to solve the triple crunch of the credit crisis, climate change and high oil prices.'[1] Ann Pettifor

---

1   Ann Pettifor, *The Case for the Green New Deal* (London & New York: Verso Books, 2019).

was among the group and argues that in developing their plan in response to the financial crisis, their 'efforts were soon eclipsed by the chaotic aftermath of the Lehman Brothers bankruptcy.'[2] The Green New Deal would have been the perfect antidote to the chaos of 2008, but the group's plan lacked the popular appeal of later versions. It was too financially technocratic and was not attached to any grassroots movement with the power to propel it into the public imagination.

## Deal or New Deal

Green. New. Deal. Those three words weren't plucked out of nowhere and strung together because they have a ring to them. The inspiration for a twenty-first century response to climate, financial and social crises came from a twentieth century response to economic depression, mass unemployment and eco-logical crisis: the New Deal of the 1930s. Pettifor explains how President Franklin D. Roosevelt embraced British economist John Maynard Keynes' proposed monetary system, shifting power over finance from private institutions to the govern-ment, to achieve the aim of full employment. The New Deal inspired the Green New Deal, demonstrating that 'every sector of life, from forestry to education to the arts to housing to elec-trification, can be transformed under the umbrella of a single, society-wide mission.'[3] Roosevelt's New Deal built highways, bridges, public buildings including schools and hospitals, airports, parks, playgrounds and athletic fields and a multi-state power system. Through the New Deal, social security and minimum wage laws were introduced, the banks were broken up,

---

2  Ibid.

3  Naomi Klein, 'Foreword', in Kate Aronoff, et al, *A Planet to Win: Why We Need a Green New Deal* (London & New York: Verso Books).

and mass tree planting and work against soil erosion revitalised regions affected by the Dust Bowl.[4] This was an unprecedented programme of state intervention through investment and policy responding to overwhelming, overlapping social, economic and environmental crises. Today's climate crisis demands greater investment and an even greater transformation with public ownership expanded to mobilise every sector of the economy in the effort towards climate justice.

As we draw inspiration from the scale of the New Deal's achievements, we must remember its biggest victories were won through the struggle of a powerful labour movement while its implementation by the government often reproduced racial segregation and disrupted Indigenous ways of life.[5] Raj Patel and Jim Goodman explain how class struggle led to the New Deal: 'In the 1910s and 1920s, strikes and labor activism in America reached a zenith. In 1919, there were 3,630 strikes involving 4,160,000 workers, or about 4 per cent of the country's total population.'[6] Organising from below created the conditions for the ruling class to concede momentous reforms to benefit the working-class. Roosevelt famously told trade union leaders that he agreed with the plans they presented to him, but that he needed them to make him do it. Governments will not initiate a Green New Deal because it is the right thing to do or even because they agree with it. They will deliver it when we give them no other choice.

Pettifor details an example of the racism of the New Deal: 'Because Southern Democrats were anxious about the threat

---

4   Naomi Klein, *On Fire: The Burning Case for a Green New Deal* (Toronto: Random House Penguin, 2019), 'Introduction: History as teacher – and warning'.

5   Aronoff et al, *Planet to Win*, Chapter 3, 'Building Freedom'.

6   Raj Patel & Jim Goodman, 'A Green New Deal for Agriculture', *Jacobin*, https://jacobinmag.com/2019/04/green-new-deal-agriculture-farm-workers (last accessed 30 March 2020).

to their racial order posed by Roosevelt's programmes, and because the president depended on their support, he bowed to their racist demands for the segregation of whites and blacks within the Civilian Conservation Corps [an environmental conservation program].[7] Black Lives Matter in the US has highlighted how racism dominates the social and economic lives of people of colour to this day. In the UK, the government's set of racist immigration policies is literally called the Hostile Environment. The Green New Deal must structure racism out of economic policy, rather than further bake it in as the New Deal did. Like anything, this requires struggle. Our lesson from the New Deal is that governments are susceptible to pressure from both reactionary forces promoting racial segregation or discrimination and progressive forces like the labour movement. Any Green New Deal proposals should make anti-racism central, for example by including those often excluded from state welfare provisions (such as, those with precarious or undocumented immigration status) and giving communities of colour real political and economic power.

## Labour's Green New Deal

In March 2019, the campaign group Labour for a Green New Deal was launched. From September 2018, I was involved in founding this new organisation as a platform for members and supporters of the UK Labour Party to organise around a shared vision for socialist climate justice. We stepped into the space created by Jeremy Corbyn's leadership of the Labour Party for members shape party policy. As the Green New Deal came to prominence in US politics, we used the framework as it captured

---

7  Pettifor, *Green New Deal*, Chapter 2, 'Public Authority and Nature's New Deal'.

the marriage of class and climate politics desperately lacking in the UK. Labour for a Green New Deal's first six months were a whirlwind. We tapped into a bubbling enthusiasm for a socialist climate politics to quickly built a volunteer organisation and network of local groups around the country, organising for Labour to adopt a socialist Green New Deal.

Over 120 Constituency Labour Parties (CLPs) submitted motions to Labour's 2019 National Conference supporting a Green New Deal. Following negotiations between activists, trade unions and party officials to composite those 120 motions into one single motion that could be debated and voted on by delegates, it was passed overwhelmingly. Just six months of grassroots organising led to the UK Labour Party supporting a Green New Deal which included a target to achieve net-zero emissions by 2030, expanding public ownership across the economy, and international resource transfers. This was a monumental commitment to rapid decarbonisation and economic transformation by a major social democratic party, and a ground-breaking case of ordinary members shaping party policy on a major crisis through grassroots organising. Entering the 2019 General Election, Labour put the spirit of the Green New Deal at the forefront of their manifesto. Though it was framed as a Green Industrial Revolution, the manifesto included the 2030 net-zero commitment to set the pace for a range of radical policies including housing insulation, transport, public ownership of energy generation and supply, and international transfers of technology to the Global South.

Labour didn't win the 2019 election. Brexit, not the climate crisis, dominated the national debate leading the Conservatives to triumph with their promise to 'get Brexit done'. Despite the loss, Labour's election campaign was probably the largest electoral mobilisation the UK has ever seen. A significant proportion of activists who campaigned by phone-banking and door-knocking were inspired to do so by Labour's socialist Green New Deal policies. By April 2020, Keir Starmer was elected to replace Jeremy Corbyn as Labour's leader. Starmer had promised to stay committed to the Green New Deal calling for it to be 'hardwired into every department at every level, from central government down to local authorities.'[8] The Green New Deal was enduringly popular, but whether the rhetoric is translated into substance is an open question. The Green New Deal is a terrain of political struggle. We must fight to defend its core tenets of mass investment, regulation and green jobs. We must also fight for a Green New Deal that goes as far as possible in its ambitions to transform the economy. Only a truly socialist Green New Deal will do, so here are the beginnings of a blueprint for one.

## 1. Democratic public ownership

The central pillar of a socialist Green New Deal is bringing as many sectors of the economy into the sphere of public ownership as possible. Where the economy is owned privately, the profit motive governs at the expense of people and planet. The private sector is incapable of mobilising to deliver an energy transition fairly or quickly. Under public ownership, including democratic oversight by workers, we can prioritise longer-term interests like climate justice and social wellbeing. We don't have to put profit before workers' rights. We can forgo profits for expensive but socially necessary work like decarbonisation. In the context of the climate crisis, a socialist Green New Deal should prioritise national public ownership (compared to municipal or community ownership) where the state nationalises existing

8   Keir Starmer, 'Internationalising the Green New Deal', https://keirstarmer. com/plans/internationalising-the-green-new-deal/ (last accessed 2 January 2021).

industry or creates new national public companies. The more of the economy that is nationally owned, the more effectively government can centrally plan the energy transition by coordinating industry which might usually be in competition. The state can provide guarantees around workers' rights and service provision that the market cannot.

Of course much of the world's industry is state owned, including fossil fuel companies (from Russia to Saudi Arabia to Venezuela), and is not being mobilised for a Green New Deal-style decarbonisation effort. Nationalised industry does not guarantee justice or decarbonisation through economic planning, but it allows for the possibility of the Green New Deal's economic mobilisation. With workplace and national economic democracy, the Green New Deal can be realised from the basis of national public ownership. Privatisation, on the other hand, is an immutable blockage.

## 2. Dismantle the fossil fuel industry

Although some versions of the Green New Deal (such as Ocasio-Cortez and Markey's resolution) do not mention fossil fuels by name, dismantling the industry is another central pillar of any Green New Deal that will achieve its aims. As long as fossil fuel companies continue extracting, decarbonisation will not happen fast enough and climate justice is impossible. It doesn't matter how much you expand the renewable energy industry if fossil fuels are still burned with emissions still driving climate change. We can draw inspiration from Bernie Sanders' Green New Deal proposed during his 2020 Presidential Campaign. He took a more confrontational approach to the fossil fuel industry. Alyssa Battistoni and Thea Rionfrancos wrote at the time: 'He proposes a ban on imports and exports of oil and gas, a ban on mountaintop mining and fracking, and a moratorium on permits to drill on public lands – all of which represent a dramatic reversal not only from Trump's much-maligned efforts to open up new public lands to drilling but from Obama's drastic expansion of domestic oil extraction.'[9] By nationalising fossil fuel companies, governments can shut-down oil wells, gas fields and coal mines as fast as possible.

## 3. Green jobs

A program of well-paid unionized green jobs for anybody that wants one is one element that binds all versions of a Green New Deal together. It resolves the key challenge of rapid energy transition by guaranteeing job security for workers in polluting industries. We should define green jobs expansively though. Green jobs are any zero-carbon jobs essential for reproducing a just society. Jobs in health and social care, education, public transport, food, manufacturing and building will all contribute to decarbonisation and act as the bedrocks of collective prosperity. By introducing economic democracy and expanding trade union rights, the Green New Deal reconfigures the relationship between labour, capital and economic transition. Rather than a vague promise of green jobs that may never be delivered, a Green New Deal would make workers the protagonists of the transition by giving them the power to direct their own economic futures, the opportunity to retrain through universal basic education, and the power to manage those green industries. With central economic planning from government directing the whole transition, specific workplaces and communities could feed-in from

9   Alyssa Battistoni & Thea Rionfrancos, 'Bernie Sanders's Green New Deal Is a Climate Plan for the Many, Not the Few' *Jacobin*, https://jacobinmag.com/2019/08/ bernie-sanders-climate-green-new-deal (last accessed 31 March 2020).

below with priorities for industrial transformation on the level of a neighbourhood, town or region. By guaranteeing that all green jobs would be tied to trade union recognition and by repealing all anti-union legislation, the Green New Deal would systematically re-empower organised labour to put workers in a position of strength against capital so that they can continue to drive the process through continued struggle.

## 4. Universal basic services

The Green New Deal is as much 'New Deal' as 'Green'. These well-paid, secure, unionised jobs across sectors are important for building a new economy. That's where Universal Basic Services come into a socialist Green New Deal. Four decades of neoliberalism have hollowed out the solidaristic functions of the state with public services, social security, funding for community initiatives and local government obliterated. Instead an ideology of self-reliance has been promoted under a free-market system where some will always be impoverished. Universal Basic Services recognise that there are some human needs too important to be left to the market and corporations to provide on the condition that they're profitable. Instead, the state should meet those needs through government agencies with the services available free at the point of use for everybody living in a country.

## 5. National Health Service

In the UK, healthcare and primary and secondary education have been archetypical services operating on this model. The UK's National Health Service (NHS) provides many with a lifeline whereas privatised healthcare systems like the US' leave people for dead or saddled with debt. The NHS is now being ravaged by successive neoliberal governments, cutting its funding and introducing market forces into the service. A socialist Green New Deal would protect the integrity and funding of those services, where they exist, while expanding them to cover all medical care for all people regardless of migration status. A Green New Deal needs a healthy people to contribute to its economic mobilisation. And there's little point stabilising the climate if we don't guarantee human dignity along the way.

## 6. National energy service

What would your life be like without a reliable supply of heat and electricity? Many people today cannot rely on a warm home and face the choice between eating and heating. Fuel poverty is an unnecessary scourge and an injustice that private energy companies perpetrate every day by pricing the poorest people out of energy while profiting from heating the planet to devastating degrees. A National Energy Service would give the government control over the generation and supply of energy to ensure it is clean, free and reliably distributed.

## 7. National housing service

The housing crisis has left many people homeless and many more subject to the exploitation of landlords happy to profit from poor quality, cramped, precarious housing. A National Housing Service would take the housing stock out of the hands of speculators and profiteers. Instead, local government would be empowered and given resources to build new council homes and buy up landlords' properties to guarantee a safe, warm home for all. Every home would be fitted with the highest-quality insulation and clean energy technology, including electric heaters and cookers, to contribute to eliminating fuel poverty and rapidly decarbonise housing.

## 8. National transport service

Although the COVID-19 pandemic inspired an upswing in digital communications by necessity, we still need to get around from place to place within and between cities, regions and countries for work and pleasure. Transport is a basic need which should not be restricted by ability to pay. A formative political experience for me as a teenager in Liverpool was campaigning against bus company Arriva when it hiked the bus fare for children. The hike meant that it would sometimes be unaffordable for poor kids travelling across the city to go to school. In response to our campaigning efforts, Arriva managers candidly told a group of teenagers that they were a capitalist company and making a profit was their priority (this was the moment I knew I was an anti-capitalist). A National Transport Service would guarantee access to transport, connect places regardless of route profitability, and facilitate investment in electrifying buses, trams, trains and cars to make the whole system zero-carbon. An integrated public transport system with a network of luxurious, high-speed, free, zero-carbon trains at the heart would be a flagship promise of a Green New Deal which improves lives while decarbonising. Working with other countries, the rail network could expand across continents to connect the whole world for free or cheap. An (inter)National Transport Service could undercut carbon-intensive domestic and short-haul aviation with cheaper and more comfortable alternatives, providing a source of green jobs across the country.

## 9. National food service

The vegan movement has pushed into the mainstream the recognition that industrial agriculture contributes significantly to
greenhouse gas emissions. Roughly in parallel, the reassertion of
neoliberalism in the aftermath of the 2008 financial crisis (often
under the guise of austerity) has plunged many people into food
poverty, with food banks proliferating. Having matured in this
era, the vegan movement is inflected with neoliberal ideology.
It valorises individual consumer choice, instead of structural
transformation, to the benefit of private companies searching
for new markets to profit from. While demand for vegan food
products increases dramatically, industrial agriculture continues
to expand, including in its contributions to emissions, land
grabs and deforestation. A National Food Service could begin
to concurrently address the interlinked crises of food insecurity, abuses of land rights, and emissions through the universal
provision of food free at the point of access.

The service could begin by universalising free school meals
and converting platforms like Deliveroo into meal delivery
services for the elderly and immobile, as Callum Cant proposes.[10]
It could expand to include regular deliveries of fresh produce
to peoples' doorsteps by scaling up community vegetable box
schemes; introducing state-funded cafes or bakeries to provide
free breakfast and lunch to all workers during the day; and
opening public canteens for communal eating in the evening.
None of this would preclude people from preparing their own
food bought from shops or eating out at restaurants, but it would
guarantee universal access to a free, healthy meal three times
a day through forms different to the stigmatised food bank. By

10   Callum Cant, *Riding for Deliveroo: Resistance in the New Economy* (Cambridge:
Polity Press, 2020), Chapter 7, 'Platform Cooperativism or Workers' Control?'.

expanding the state's capacity to provide food, a National Food Service expands the state's procurement power and allows for greater central planning of food both imported and produced domestically. It can use this to shift aggregate demand in favour of sustainable local, seasonal and vegetarian food, and against the most environmentally harmful products. The National Food Service could restructure our food system through procurement and planning, rather than relying on individuals to make the right choice in a system that pushes unhealthy, unsustainable food because it profits private producers.

## 10. International institutions

We can imagine these universal basic services within our own nations' borders, but the climate, economic and social crises the Green New Deal seeks to address are global in scale. A socialist Green New Deal should not reproduce the colonialism of the system that plunged us into climate crisis with luxury at home and poverty abroad. In response to the domestic focus of some iterations, some climate justice activists have called for a 'global Green New Deal', demanding 'a global response', supporting workers around the world, providing climate reparations and supply chain justice.[11] Proposals put forward by Ann Pettifor and Labour for a Green New Deal are good examples of outward-facing and internationalist interpretations of the Green New Deal while remaining chiefly national projects. The nation-state is the primary level of political power as well as the most

---

11    Asad Rehman, 'A Green New Deal must deliver global justice', *Red Pepper*, https://redpepper.org.uk/a-green-new-deal-must-deliver-global-justice/ (last accessed 10 January 2021). Andrew Taylor & Harpreet Kaur Paul, 'A "Green New Deal" needs to be global, not local', openDemocracy, https://opendemocracy.net/en/opendemocracyuk/a-green-new-deal-needs-to-be-global-not-local/ (last accessed 21 January 2021).

consistent site of democratic oversight. This is the reality of global politics today and a Green New Deal should reflect that both in its starting point and transformative ambitions. It makes sense for investment, regulation and the expansion of public ownership to begin at a national level.

The challenge, then, is how to globalise the transformations of a Green New Deal from this starting point. Powerful states like the US and UK can begin by building global coalitions to reform the institutional architecture of global governance and finance to equalise geopolitical power, uplifting countries, particularly in the Global South, that have contributed least but been hit hardest by climate and economic crises. They could construct institutions in parallel to, and eventually supplanting, the likes of the EU, IMF, World Bank and WTO with human rights, freedom, equity and ecological justice as foundational values, rather than protecting free markets and the interests of colonial-capitalist states. This new architecture could create the political and financial conditions for all nations to invest in their own just transitions and adaptation, while building the foundations of sustainable economic prosperity domestically. Failing the political conditions for multiple powerful states to begin such a project, individual states implementing a Green New Deal can begin by transferring financial, human, and technological resources to countries in the Global South keen to get on with transition. The Labour Party promised to do this in the 2019 General Election with a policy to make technology developed as part of its Green Industrial Revolution available for free or cheap to countries in the Global South.

## 11. Supply chain justice

The pace and scale of the Green New Deal's economic transformation is driven by the need to mitigate and adapt to the harshest impacts of climate change which are overwhelmingly hitting those in the Global South the hardest. If the transition does not break with business-as-usual economies of colonial resource extraction, it further threatens the poorest globally. Campaigners like Asad Rehman, director of global justice charity War on Want, has warned:

> In this new energy revolution, it is cobalt, lithium, silver and copper that will replace oil, gas and coal as the new frontline of our corporate destruction. The metals and minerals needed to build our wind turbines, our solar panels and electric batteries will be ripped out of the earth so that the UK continues to enjoy 'lifeboat ethics': temporary sustainability to save us, but at the cost of the poor.[12]

The demand for resources needed for the green energy transition have also fuelled conflict, eroded the resilience of local environments, contaminated water supplies and been monopolized by western corporations. None of these are necessary consequences of extracting mineral resources, but they become inevitable when extraction is done according to capitalist logics in the relentless pursuit of growing profits. We can imagine a world where mineral resources are treated as a public common, extracted safely with the free, prior and informed consent of communities whose land is affected. Building that world from within our unjust system as part of the struggle for climate justice is one of the great challenges for the Green New Deal's architects.

---

12   Asad Rehman, 'The "green new deal" supported by Ocasio-Cortez and Corbyn is just a new form of colonialism', *Independent*, https://independent.co.uk/voices/green-new-deal-alexandria-ocasio-cortez-corbyn-colonialism-climate-change-a8899876.html (last accessed 1 April 2020).

In his book, *Fully Automated Luxury Communism*, Aaron Bastani argues that asteroid mining could solve the problem of resource scarcity. Minerals like cobalt and lithium are abundant on asteroids hurtling through space. We could cut out the injustices associated with mining those resources if we could master the technology to land spacecraft on asteroids, mine them, and bring the resources back to Earth. Bastani doesn't just think this is possible, but inevitable.[13] The question is whether capitalists like Elon Musk profit from the abundant resources of infinite space, or whether we treat space as a global common to be used for the public good like providing the resource basis of a Green New Deal. It is unlikely that asteroid mining will be mastered in the coming decade, and the technology alone does not address the root causes of the injustices of extraction under capitalism. But a Green New Deal could include investment and research into developing asteroid mining as a long-term solution to demand for scarce mineral resources, and an effort to keep space exploration in the public sector for the common good.

In the immediate term, a socialist Green New Deal needs measures to begin promoting justice in global extraction. Firstly, using state procurement power (expanded with greater public ownership across the economy) to promote labour and environmental standards in supply chains. Secondly, using new international institutions to promote the democratic ownership of resources by states with strong checks and balances guaranteeing resource sovereignty for local or Indigenous communities. Third, reducing domestic aggregate demand for energy (in turn reducing demand for such resources) through insulation of buildings, smart grids and the collective pooling of energy demand through the expansion of public ownership

---

13 Aaron Bastani, *Fully Automated Luxury Communism: A Manifesto* (London & New York: Verso Book, 2019), p. 134.

and services. Fourth, investing in the development of synthetic minerals where possible. Fifth, extracting resources domestically or in the Global North where they exist as priority for consumption in those countries. The challenges of supply chain justice are certainly among the most difficult to resolve through the Green New Deal, but rather than cower at their magnitude, we should embrace them and commit to being as bold as possible in the measures we propose. Because half measures will never be enough.

I have outlined a blueprint for a socialist Green New Deal, including the elements that should be at the core of any serious economic mobilisation in the coming decades of compounding crises: public ownership, green jobs, just transition, universal basic services and global justice. One of the joys of the Green New Deal is that this blueprint need not be fixed. Like the New Deal of the 1930s, it is a framework through which to respond to the major crises of our time. It is an opportunity, emerging from a decade of austerity and over four decades of neoliberalism, to collectively take risks, experiment and eventually discover the initiatives that will scale to become the basis of our new economy and a just society.

# Chapter 6
# Jobs, jobs, jobs

Trade unions are among history's most powerful and enduring vehicles for furthering the collective interests of working-class people. As climate change hits the poorest hardest, we need a powerful trade union movement leading the struggle for a Green New Deal so that workers are the protagonists, not victims, of economic transition. Workers are uniquely placed in the economy to lead such a movement. By withdrawing their labour or disrupting key sites of economic production, workers have the power to extract demands from capital and governments and lay the foundations for a new economy. Right now, many in the trade union movement are not prepared to play this role. After decades of brutal assault by capital with wave after wave of neoliberal reforms, trade unions are generally lacking in membership, power, rights and victories. But trade unions have survived this onslaught and continue to provide an organisational vehicle for workers to collectively protect their interests. With the climate crisis upon us, the task of rebuilding the power and confidence of the labour movement is urgent.

The Green New Deal framework unifies climate and class politics, but in recent decades many have perceived a conflict between them. Environmentalists have argued for blocking the expansion of high-emissions infrastructure at all costs, including airport runways, roads and coal mines. In the UK, trade unions like GMB and Unite have argued that, while they support the principles of a just transition, their priority is safeguarding the existing jobs of members.[1] Each of these positions assumes that only incremental gains can be won within the status quo. All the while, our economic system remains intact, emissions rise, and workers are still exploited. The interests of workers and those on the sharp end of climate injustice only appear in conflict if we assume the permanence of business as usual. Capitalism is reliant in equal measure on exploiting workers' labour and turning nature into commodities, regardless of the socially and ecologically devastating side effects. The climate and labour movements can unite around this shared antagonism with capital and in support of a transformational Green New Deal for economic and climate justice.

## Power in a union

In his book, *Why You Should Be a Trade Unionist*, Len McCluskey, General Secretary of Unite the Union, tells the story of trade unions' past as powerful mass membership organisations responsible for advances in working conditions enjoyed today: shorter working hours, higher wages, health & safety standards and trade union recognition.[2] These victories have come from a repertoire of tactics which leverage workers' collective power against the bosses. We call this industrial action. Chief among these tactics is the strike, where workers collectively refuse to work. Other tactics include occupying the workplace, work slowdowns, workers simultaneously taking sick leave, working

---

1   Both are general unions representing workers across the economy, particularly in the private sector, with memberships of around 620,000 and 1.2 million respectively.
2   Len McCluskey, *Why You Should Be a Trade Unionist* (London & New York: Verso Books, 2019), p. 15.

specifically to the hours in contracts or working specifically to the letter of the contract.

These tactics have been used successfully as long as workers have organised together. The first recorded strike took place in Egypt in 1152 BC as artisans and tomb workers organised an uprising demanding pay. Trade unions became common with industrial society in Britain and its colonies from the 1600s onwards.[3] Howard Zinn describes strikes in the colonial United States from the 1640s by fishermen, carpenters, coopers, butchers and bakers against withheld pay, inadequate food, sackings, government fees and high prices.[4] Eventually, trade unions were legalised in Britain in 1824. By the late 1800s, a 'new unionism' emerged where 'unskilled' and low-paid workers began to organise. In 1888, working-class women and teenage girls working at the Bryant & May match factory in London went on strike against very low wages, physical and mental abuse by bosses, and exposure to lethal white phosphorous. They shut the factory down for sixteen days and won improvements including trade union recognition.[5] In 1889, new unionism escalated in scale as London dockers went on strike, winning a pay rise for thousands and forming a union.[6] Trade union membership, and therefore pay and conditions, grew during the First World War. In 1917, a series of strikes across the economy played a key role in the Russian Revolution. Unions were strong in the decades after the Second World War. In 1974, striking miners demanded

---

3   Henry Pelling, *A History of British Trade Unionism* (Basingstoke: Palgrave Macmillan, 1992), pp. 10–23.

4   Howard Zinn, *A People's History of the United States* (New York: Harper Perennial Modern Classics, 2015), p. 52.

5   Louise Raw, *Striking a Light: The Bryant and May Matchwomen and their Place in History* (London: Continuum, 2009).

6   Pelling, *A History of British Trade Unionism*, pp. 94–7.

better pay and effectively brought down Edward Heath's Tory government, resulting in the election of a Labour government.[7]

However, since Margaret Thatcher's election as Prime Minister in 1979, government after government have continued her assault on the power of trade unions in the UK, from the miners' strike from 1984–5, a historic defeat of the labour movement by Thatcher's government, to David Cameron's Tory government introducing the Trade Union Act 2016 making balloting for strike action even harder. For strike action to be legal, over 50% of workers must vote in a ballot and a majority must be in favour.

As part of the Green New Deal's economic transformation, high-emissions industries will contract, and the likes of fossil fuel production will become obsolete. For this transition to be just, new industries must be introduced in their place with economic democracy and trade union recognition so that workers direct their own futures. Just transition is not inevitable, though. Where workers do not lead the transition, trade unions may use industrial action to resist these changes and protect workers' immediate interests. For example, trade unions in Poland have taken strike action demanding guarantees for the future of coal as part of the country's national energy plan.[8] Working in the fossil fuel sector right now is precarious. Demand for coal is plummeting, Poland is under pressure from the European Union to phase it out, but there is no clear alternative. Though trade unions have the potential as vehicles for class struggle towards economic transformation, in cases like this they act to preserve

---

7   McCluskey, *Be a Trade Unionist*, p. 23.

8   'Miners at Poland's biggest coal group protest over pay, energy plans', *Reuters*, https://reuters.com/article/poland-miners-strike/miners-at-polands-biggest-coal-group-protest-over-pay-energy-plans-idUSL8N2AH1MA (last accessed 9 April 2020).

the status quo. When they primarily conceive of themselves as negotiating the best conditions for workers under capitalism, the union's interest becomes maintaining things broadly as they are, seeking only marginal gains.

On the other hand, history gives us examples of unionised workers proactively fighting for a just transition to a new economy. The Lucas Plan is the most famous example. In 1976, workers at the Lucas Aerospace factory (a weapons manufacturer) faced structural unemployment as technological changes automated some out of their jobs. The Lucas Aerospace Combine Shop Stewards Committee developed an alternative proposal to maintain jobs by running the company for socially useful ends. They assessed the skills and equipment at their disposal and produced a 1,200-page plan including proposals for 150 products covering 'medical technologies, renewable energy, safety improvements and heating technology for social housing.'[9] The plan was eventually rejected by the Labour government of the time and wasn't supported by national trade unions. But the Lucas Plan offers an example of workers autonomously organising to use their skills and resources to contribute to the transformation of economic production. It's an inspiration for the economic democracy central to a Green New Deal and the possibility of trade unions organising for alternative industrial strategies.

## Heathrow expansion

The debate over Heathrow Airport expansion is one of British politics' longest running sagas. Trade unions and environmentalists have been embroiled as central protagonists in the plot.

---

9   Hilary Wainwright and Dave Elliot, *The Lucas Plan: A New Trade Unionism in the Making?* (London: Allison & Busby, 1981), pp. 101–7.

Along with former Prime Minister Gordon Brown, Unite, GMB and the Trades Union Congress have all backed the project. Boris Johnson opposed Heathrow expansion while he was Mayor of London and later MP for a nearby constituency along with other local MPs. Johnson's reasoning was opportunistic, but he was joined by environmental NGOs like Friends of the Earth, whose criticism have focused on the increase in emissions that aviation expansion would bring. Aviation accounts for around 3.5 per cent of emissions globally, though a WWF report ranks air travel accounting for 5.9 per cent in the UK.[10] This is a significant minority, but airport expansion is a particularly hot issue in decarbonisation debates. Unlike other high-emissions sectors like heating homes, car fuel or electricity, it is not clear that the aviation industry can actually be decarbonised (certainly not on the timescale the climate crisis commands), such is the fuel required to get a large plane into the sky. If Heathrow is expanded, it locks the economy into greater emissions for the foreseeable future as more planes fill the extra runway, burning more fuel in the process, to guarantee sufficient returns on the investment.

Justifying Unite's support for Heathrow expansion despite the emissions and local pollution, Len McCluskey argues: 'But these are good, skilled jobs, and in the absence of a government with the sort of radical vision that Labour has for alternatives I am bound to support members whose families and communities depend on those jobs.'[11] McCluskey implicitly accepts the permeance of a status quo so rigid that not even an alternative

---

10  WWF-UK, 'Carbon Footprint: Exploring the UK's Contribution to Climate Change, March 2020, https://wwf.org.uk/sites/default/files/2020-04/FINAL-WWF-UK_Carbon_Footprint_Analysis_Report_March_2020%20%28003%29.pdf  (last accessed 25 March 2021).

11  McCluskey, *Be a Trade Unionist*, p. 130.

industrial strategy is possible without a radical Labour government. Trade unions are apparently doomed to welcoming a modest set of good, skilled jobs to marginally increase union membership, in exchange for infrastructure guaranteeing social and environmental injustices for other working-class people. Why should electing a Labour government be the endpoint of the trade union movement's strategy? The largest trade unions individually have higher memberships than the Labour Party and are well-resourced organisations with active branches spread across the country. Unions have the power and ingenuity of organised workers to develop alternative plans and strike to win them. Throughout its history, the labour movement has won and lost under both Conservative and Labour governments. While Labour governments are certainly preferable for the labour movement, the Conservative Party has dominated government over the last century. We need a strategy of winning when our enemies are in charge, forcing them to pull the levers of state power in a progressive direction. It's true that a program on the scale of a Green New Deal will have to be implemented by a progressive government, but until then trade unions can join social movements in creating the political conditions for such a transformation by winning on alternative infrastructure projects and the building blocks of a just transition.

That Heathrow expansion is still on the table is as much the fault of environmentalists opposing it as trade unions supporting it. Grassroots direct-action groups and NGOs have since 2007 successfully delayed Heathrow expansion, but never had the project decisively scrapped. As MPs voted on the plan in the House of Commons in June 2018, the Conservative government whipped its MPs to vote in favour while Labour officially opposed expansion but gave its MPs a free vote. Many voted with the government and expansion was approved with a large majority of 296 votes. By February 2020, Heathrow expansion was ruled illegal by the court of appeal after a legal challenge led by Friends of the Earth and legal charity Plan B. The court ruled that expansion did not consider the government's commitments to address climate change so could not proceed. In December 2020, the Supreme Court overturned the February judgement, leaving Heathrow free to seek planning permission for the expansion. Relying on the courts is a precarious position for campaigners to be in as legal systems in capitalist economies tend to be weighted towards protecting the interests of capital.

Those campaigning against Heathrow expansion should never have allowed the battle over Heathrow get to a stage where MPs felt emboldened to vote in favour, leaving the expansion's future in the hands of the courts, beyond democratic oversight. As positions hardened around Heathrow expansion, environmentalists have often assigned blame to trade unions like Unite. I've heard anti-aviation activists argue that trade unions are as responsible for climate change as multinational corporations, blame Unite for Labour's policy failings, and outright dismiss workers' concerns around job security as unimportant compared to environmental issues. Direct-action groups and NGOs have spent years petitioning, lobbying, litigating and blockading against Heathrow expansion. At no point did they seek to build solidarity with trade unions or workers involved. They did not seek to understand and transform the structural constraints the labour movement operates within. They did not offer time and resources to co-develop alternative industrial plans to airport expansion and campaign to demand them together.

I spoke to Gabrielle Jeliazkov, a just transition campaigner with Platform London. I asked her how the next decade will affect workers in another high-carbon industry, North Sea oil and gas:

In the North Sea specifically something like 70% of the workforce are contractors, so what you're looking at is a workforce that has weak employment rights, little funded training from their companies, and little protection when it comes to their employment status. As the climate crisis continues and as the phase out of oil and gas becomes even more urgent, we're just going to see even more escalating inequality and insecurity in the workforce. Without massive mobilization on the part of the workforce we're facing a very bad situation for oil and gas workers.

What are the barriers to unions taking more of a leadership position within the climate movement?

A barrier to unions being centred in climate work is when the climate movement does not take a working-class, anti-racist, accessible approach to climate change. The things that people need are secure housing, food security, sustainable jobs, and if the climate movement is built on these things – if those are things we're fighting for – then everyone will fight for that. Someone working 50 hours a week on an insecure contract isn't going to spend their few hours not working fighting for the relatively abstract goal of staying within 1.5 degrees. They're going to fight for a job, housing, and food security. And if we are fighting for climate justice, then we are necessarily fighting for all of those things, we are fighting for redistribution and against rising inequality in all senses of the word.

What should the climate movement's attitude to unions be?

The climate movement has to speak about things that people need. And if we want to push our demands forward, then that means necessarily that the demands need to be built by the people who are going to be most impacted by them, and that's quite an uncomfortable thing for the climate NGOs and less diverse climate movement bodies because it means their strategies have to change, but that's what solidarity is.

## A context of defeat

While some environmentalists chastise trade unions like Unite for their position on specific infrastructure projects, anyone committed to building climate justice should begin by understanding that 'trade unions operate in a context of defeats.'[12] Mainstream trade unions have been forced to retreat into a strategy of service provision for members, while collaborating with capital to achieve marginal gains. McCluskey explains:

> The vast majority of trade unionists' time is spent dealing positively with employers over pay and conditions and resolving grievances, and I am proud of the type of engagement we have with them. I talk to company chief executives every week, often working together to resolve complex issues that as an employer they cannot solve on their own. That is what unions and their shop stewards do – they negotiate and work to put out fires. We do not seek confrontation, and we do not relish fights; but neither do we walk away from bullying bosses and companies that are not treating employees fairly. So when we do take industrial action, it is a last resort.[13]

McCluskey begins his book by recounting the history of material gains won by trade unions through industrial action and antago-

---

12    Jamie Woodcock, *Working the Phones: Control and Resistance in Call Centres* (London: Pluto Press, 2017), p. 139.

13    McCluskey, *Be a Trade Unionist*, p. 27.

nism with the capitalist class, before making the case for today's conservative settlement where trade unions work with bosses as problem solvers. Industrial action is only a last resort, but perhaps this is because the established labour movement has forgotten what it's like to win big. Where trade unions do push back against business as usual, Nick Srnicek and Alex Williams identify a tendency towards defensive demands to 'save our health system' or 'stop austerity', representing this attitude to the status quo where 'the best we can hope for is small impediments in the face of predatory capitalism.'[14]

Someone in Len McCluskey's position might argue that as General Secretary of a major trade union, operating in this context of historic defeat, his role is to steward the union with a pragmatic strategy of survival. On the contrary, as we face up to the scale of the climate crisis and the injustices it delivers for workers and the global poor, trade unions must break with this conservative disposition. Instead, unions should embrace radical politics, demands and strategies for climate justice through economic transformation. Capital has waged an all-out war on labour for decades, with climate change a cruel side effect. Through their trade unions, it's time for workers to fight back and build a new economy.

## Striking to win

The conservatism of larger trade unions is not reflected evenly across the labour movement. In the UK, some of the most encouraging worker organising in recent years has come against food delivery platforms like Deliveroo and Uber Eats. Callum Cant has written of how 'a strike by Deliveroo workers in London in

---

14    Nick Srnicek & Alex Williams, *Inventing the Future* (London & New York: Verso Books, 2016), p. 47.

the summer of 2016 was the first sign that food delivery platform workers were capable of mass collective action.'[15] Riders did not have direct employment status, which both meant they lacked important rights but were also able to autonomously organise a series of wildcat strikes free from the 'state repression of strikes' and bureaucratic trade unions. The strikes quickly spread across the UK and Europe, later to South Africa and China. Following three nationally coordinated strikes in the first six months of 2019, Woodcock and Cant write that in the time between the first strike in 2016 and now, the struggle of platform workers has developed from 'a workforce finding its feet, but of a workforce that has begun to wage a widespread, intense, and prolonged struggle.'[16] A workforce once studied for its supposed inability to be organised has become one of the brightest sparks in the global labour movement.

In the US, teachers have led the labour movement by taking strike action in multiple states. Jane McAlevey tells the story of the 2012 Chicago Teachers' Strike in her book *No Shortcuts*. Militant rank-and-file members organised in internal elections to take control of the Chicago Teachers' Union, including electing Karen Lewis, a local black and Jewish teacher of over 20 years, as president. The new leadership organised the union, educated its members and prepared to take action ahead of contract negotiations with Chicago mayor, Rahm Emanuel. The teachers were on strike from 10 to 18 September 2012, grinding the city to a halt, winning the support of parents, children and the city as a whole. They won a pay rise, defeated Emanuel's

---

15   Callum Cant, 'Precarious couriers are leading the struggle against platform capitalism', Krytyka Polityczna & European Alternatives, http://politicalcritique. org/world/2017/precarious-couriers-are-leading-the-struggle-against-platform-capitalism/ (last accessed 10 April 2020).

16   Jamie Woodcock & Callum Cant, 'The End of the Beginning', *Notes from Below*, https://notesfrombelow.org/article/end-beginning (last accessed 10 April 2020).

attempt to introduce performance-related pay and abolish tenure for teachers. McAlevey notes that the strike was indeed a defensive one, but 'Chicago's teachers have proven that a broken union can be rebuilt in a *very* short time – less than two years.'[17] Inspired by the Chicago teachers' strikes and instigated by young teachers who had cut their teeth in the Bernie Sanders for President campaign in 2016 was a wave of teachers strikes in Republican-dominated states beginning in 2018. In West Virginia, Arizona and Oklahoma, thousands of teachers went out on strike 'for better pay and school funding' resulting in 'a range of major victories.'[18] They won pay rises, defeated anti-union legislation, stopped new tax cuts and won new school funding. This was all in states where restrictive union laws meant teachers taking strike action was illegal.[19]

Eric Blanc explains in his book *Red State Revolt* that the weakness of unions in these states meant 'the strikes took on an unusually volcanic and unruly form.' As with the transnational platform strikes, without an overbearing union bureaucracy, rank-and-file members 'stepped into the leadership vacuum and filled it to the best of their abilities.'[20] Over the last decade, organised workers have made the most explosive impact where they've faced the toughest conditions: no direct employment, weak unions and anti-union legislation. Taking industrial action in the age of climate change means confronting those conditions while directing action towards not just individual bosses, but a crisis-ridden economic system.

---

17  Jane McAlevey, *No Shortcuts: Organizing for Power in the New Gilded Age* (New York: Oxford University Press, 2016), p. 142.

18  Eric Blanc, *Red State Revolt* (London & New York: Verso Books, 2019), Introduction.

19  Ibid., 'A Historic Upsurge'.

20  Ibid.

In the UK, the defeat of the miners' strikes and neoliberal reforms have had a chilling effect on industrial action. However, militant unions have still been able to win for their workers. For example, the National Union of Rail, Maritime and Transport Workers (RMT), has been led by communists and socialists, repeatedly using strikes to improve pay, conditions, pensions and delay privatisations. Between 2000 and 2008, RMT balloted for strike action 50 times for the London Underground and 68 times on the railway network, resulting in 18 and 33 strikes respectively.[21] Whereas other unions have retreated into a conservative approach, often despite socialist leadership (as with McCluskey's Unite), the RMT's socialist leadership, combined with the leverage of taking action in the industrially significant transport sector, has cultivated a powerful militancy in the face of neoliberal privatisations and class warfare against workers.[22]

We need to get to a place where workers collaborate across sectors and across movements, taking action for our shared demands. Imagine workers going on strike in refusal to build emissions-intensive infrastructure like new airports, while working to demand investment to expand basic services and grow the country's zero-carbon rail links. In August 2019, workers in Belfast took strike action in that spirit. They were employed at the Harland & Wolff shipyard (which built the Titanic) as it went into administration, because its Norwegian parent company couldn't find a buyer. Instead of rolling over and accepting their looming unemployment, workers occupied the shipyard and went on strike for three weeks blocking insolvency practitioners from entering the site. The workers' demands were for the docks to be nationalised (to save them from closure) and

---

21    Ralph Darlington, 'Leadership and union militancy: The case of the RMT', *Capital & Class*, 99 (2002), p. 6.

22    Darlington, 'Leadership and union militancy, pp. 14–15.

to be used to produce renewable energy infrastructure. They had previously built parts for wind turbines, so this was a 'not only [. . .] a sustainable solution, but also a practical one because of their skill set.'[23] By October, jobs on the site were saved, as a buyer was found. But this meant the workers' demands for nationalisation and transition weren't met. Regardless, Harland & Wolff workers led by example in taking industrial action for climate justice.

Around the same time in the US, in September 2019, workers at the world's biggest technology giants took action for climate justice too. Amazon Employees for Climate Justice joined similar groups at Google and Microsoft in taking action with work stoppages on the same day as global youth climate strikes. Workers across the companies had the same set of demands: zero emissions by 2030; zero contracts with fossil fuel companies; and zero funding for climate denying lobbyists and politicians.[24] The first demand implies a transformation of the economy. If Amazon, Google and Microsoft go zero-carbon, swathes of the economy will follow. The second and third look to pit tech capital against fossil capital, undermining the grip of Big Polluters on politics and technological development. Though the latter two are more conceivable within the constraints of the status quo, they do all imply the de-prioritisation of the profit motive in favour of decarbonisation. Given how expensive rapid decarbonisation would be for these tech companies, and failing bold government intervention, only militant industrial action

---

23 Lauren Kaori Gurley, 'Workers Seize the Shipyard That Built the Titanic, Plan to Make Renewable Energy There', *Vice*, https://vice.com/en_uk/article/8xwanz/workers-seize-the-shipyard-that-built-the-titanic-to-make-renewable-energy (last accessed 15 April 2020).

24 Google Workers for Climate Action, 'Google Workers are Striking for Climate on Sept 20', *Medium*, https://medium.com/@googworkersac/google-workers-are-striking-for-climate-sept-20-7eba2100b621 (last accessed 15 April 2020).

paralysing the companies' profits to a greater extent can force those changes. Amazon workers have highlighted the disproportionate leverage that they have over the company, as primarily software engineers or project managers who are in high demand in the labour market. The Tech Workers Coalition in the US is one example of attempts to build workers' power in the sector. If they can successfully scale up in confidence and organisation to take serious, protracted strike action, they stand the best chance of anyone to force capital to decarbonise on the timescale necessary.

Workers in different sectors will all have something unique to contribute, though. Eric Blanc uses the Red State teachers' strikes to explore the impact public sector strikes can have despite not directly diminishing private sector profits. He argues that public sector strikes 'win by creating a social and political crisis.' Climate justice demands a transformation of the entire economy, so forcing political crisis in the public sector and damaging profitability in the private sector can form complementary strands of a confrontational strategy by workers to demand just transition and adaptation.

One lesson we can take from transnational delivery riders, teachers, transport workers, construction workers and organised labour throughout history is that when unions take militant action and win, they grow their power, membership, and win some more. The climate crisis has made class struggle in the workplace, in national politics and internationally, more urgent than ever. A labour movement organising for climate justice will need a range of approaches to a variety of contexts: old and new unions, strong union rights and anti-union laws, secure and precarious work, private and public sector, transition industries and those of the future. In all of the above examples, workers are organising and winning in challenging circumstances. In some of these cases, workers are fighting for their own interests with a complementary focus on the wider political economy: the RMT supports public ownership, Harland & Wolff workers supporting just transition, tech workers prioritising decarbonisation.

In all of these cases, however, there still lacks a strategy for holistic economic transformation beyond the workplace. The RMT, for example, is a militant union led by socialists, but has not actively brought the fight for public ownership. To get serious about climate justice, we need to level up unions' confidence and ambition to take industrial action to expand public ownership, demanding green infrastructure and industries, and blocking carbon-intensive infrastructure. In some contexts, workers will need to form new unions and take wildcat strike action. We can certainly draw inspiration from where this has happened in recent years. To scale a new industrial and political militancy for climate justice, workers will generally need to use existing union structures. So when you hear regular calls to 'join a union', you should absolutely do so. But don't leave it at that. Join a union and agitate for deeper member democracy, accountability for the leadership, and for a fighting strategy of industrial action for climate justice. We can take inspiration from Chicago's teachers capturing their union and instilling a culture of militant organising. For a Green New Deal including just transition, economic democracy and expanded public ownership, we need a renaissance and revolution in the labour movement. It's up to rank-and-file workers to make that a reality.

# Chapter 7

# The s-word (state power)

From 2014 onwards, I was involved in a number of campaigns that aimed to confront the fossil fuel industry and that had the end goal of totally dismantling it. In the anti-fracking movement, we targeted the small number of exploratory drilling sites, aiming to undermine their viability and stop this new industry from establishing itself in the UK. In the fossil fuel divestment movement, we aimed to lay the foundations of political action against the industry by stigmatising it and marginalising fossil capital from social and political life. In climate finance campaigns, we aimed to drive a wedge between banks and fossil fuel companies by starving the latter of the finance they needed to carry on extracting. In all of these examples, the campaigns made commendable gains on their own terms, while leaving the fossil fuel industry relatively stable. There has still been no commercial fracking in the UK, but for every direct-action campaign, dozens more fossil fuel projects around the world pop up without opposition. Trillions of dollars have been divested from fossil fuel companies, but they continue to expand operations, and there is an absence of strong political action to dismantle the industry. Banks have set net-zero targets and some now exclude the most polluting fossil fuels (like coal, tar sands and Arctic drilling), but have often found loopholes to grab a headline while continuing to pump cash into the industry.

Having been involved in these campaigns for some years, I started to reflect on why they were not taking us towards a just transition out of fossil fuels on the timescale we required. In each case, it was clear that it would require an intervention from the state to realise the end goal of the campaigns. For the anti-fracking movement to win, the government would have to ban fracking and other new fossil fuel exploration. For banks to completely end finance for fossil fuels, the government would need to introduce strict regulation and heavy penalties for non-compliance. For divestment to be at all worthwhile, governments would have to step into the political space divestment has created and enact material measures to dismantle the industry. The problem, of course, is that most governments are unwilling to go that far. Most climate campaigns are reluctant to adopt strategies of directing state capacity for our ends, let alone capturing state power ourselves. If we are serious about climate justice, this has to change.

## Climate anarchisms

There are roughly three areas of thought (and practice) around the climate that share a similar attitude, though for different reasons, that overall the state does more harm than good. This shared attitude informs their respective approaches to climate action, which minimise the role of governments and the importance of state capacity. The three tendencies are the private sector, the localist or co-operative sector, and the radical climate movement.[1]

---

1   By 'radical climate movement' here I tend to mean direct-action networks and some of the more justice-oriented NGOs: the desire to dismantle the fossil fuel industry is a unifying aim.

The private sector's approach to decarbonisation has been to enthusiastically support the vague, non-binding and distant targets agreed at international conferences (UNFCCC conferences are often sponsored by an array of the most polluting corporations),[2] while lobbying against legislation that would tax polluting activities or seriously mandate companies to make meaningful changes.[3] Fossil fuel companies and banks have even made their own commitments for their business to reach net-zero by 2050, but rarely include detailed plans of how to get there and often exclude core polluting activities like extraction. The private sector will outwardly say that it will adapt to climate change best if left alone, free from the market-warping constraints of government innovation. It will claim that innovation is best left to profit-driven enterprise and that the public sector is poorly managed, inefficient, and starved of investment.

The co-operative movement and local initiatives embody the principle that 'small is beautiful'. Co-operative models of housing, food distribution, and renewable energy production seek to give workers and communities direct control over those essential sectors of the economy. The approach emerges from a criticism of politics on a national scale – whether socialist or capitalist – as authoritarian and anti-democratic. The more centrally the economy is managed, the less say that those who actually experience it have in its management. Calls to go 'back to the land' or 'grow your own' food emphasise a belief about

2   Caitlin Tilley, 'UK Government's COP26 Sponsorship Choice "Nowhere Near Good Enough"', Campaigners Say', *DeSmog UK*, https://desmog.co.uk/2020/11/19/uk-government-s-cop26-sponsorship-choice-nowhere-near-good-enough-campaigners-say (last accessed 30 January 2021).
3   Sandra Laville, 'Top oil firms spending millions lobbying to block climate change policies, says report', *Guardian*, https://theguardian.com/business/2019/mar/22/top-oil-firms-spending-millions-lobbying-to-block-climate-change-policies-says-report (accessed 24 January 2021).

scale in agriculture. The idea is that large-scale industrial agriculture, with its mono-culture crops and global transportation of goods, is to blame for the polluting excesses of our capitalist food system. The solution, then, is to return to local and ecologically sustainable models of farming.

The radical climate movement contains an often-incoherent mixture of political orientations, with more or less explicit attitudes to state power. Its unifying belief is that if we build strong enough people power, we can beat the fossil fuel industry despite corporate power and government inaction. Many who participate in direct-action networks (like Reclaim the Power or Ende Gelände) are likely to have an explicitly anarchist politics, entirely rejecting the state as an agent of progressive change. The state is perceived as inherently violent; whoever is in power, its primary function is the oppression of women, people of colour, migrants, the poor and other marginalised identities through policing, incarceration, military and government. These beliefs inform direct-action strategies where activists feel they have to take matters into their own hands: 'the government won't confront fossil capital, so we have to put our bodies in the way to stop extraction'. Radical NGOs and campaigns generally adopt a less explicitly anarchist outlook, toning down the open hostility to the state, but are informed by similar logics. The prioritisation of campaigns for institutions (universities, local government and philanthropists) to divest and finance (banks and insurance companies) to defund fossil fuels is instructive. In seeking only to divert flows of capital, neglecting strategies to leverage state power to dismantle the fossil fuel industry and transform ownership in the economy, the radical climate movement implicitly accepts the permanence of the neoliberal status quo where the private sector is left alone by the state.

## State capacity

In each of these climate anarchisms, there is some useful criticism of the state which we should learn from. The private sector's caution of overly bureaucratised public sectors with chronic under investment isn't wholly misplaced. Localism and co-operative movements are keen to safeguard economic democracy and control. Radical climate movements are healthily sceptical of the violence of the modern state. However, none of these criticisms are so strong that they cannot be mitigated for. In fact, if we are to deliver climate justice it will be entirely necessary to do so through the institutions of nation states. The state is hardly a perfect political formation, but it is the most appropriate we have right now for addressing the climate crisis with its scale, urgency and complexity.

What we mean by 'the state' is a source of debate between political scientists. There isn't a single definition on which everybody agrees. For the sake of expediency (and hopefully with limited controversy), I'm using 'the state' to mean a set of interconnected institutions which govern a sovereign nation. Many of these institutions primarily enact the violent and oppressive functions of the state, as radical movements rightly point out. Police and the justice system uphold capitalist property rights and disproportionately incarcerate people of colour and the poor in prisons. Detention and deportation systems lock-up migrants and forcibly send them to places they may never have been or where they face violence. Militaries use deadly force to fight wars to further capital's interests and occupy foreign territories to annex natural resources and create new markets. Government departments penalise the unemployed, deprive the vulnerable of vital services, and finance fossil fuel extraction at home and abroad. As well as the bad things, we should also understand that many of the state's institutions perform socially useful functions: public education, universal health-care, research and innovation, council housing, welfare benefits, public transport and investment in infrastructure. Nation states have generally been developed over the past centuries to uphold capitalist and white supremacist interests. In all but a few cases globally, it is still capital (via its political representatives) which controls the state. Those socially useful elements of the state have been the result of powerful struggle from social and trade union movements throughout history. These concessions to the working-class demonstrate that, like most political terrains, states are fiercely contested. Although they still operate within global capitalism (and regardless of how one judges their achievements), examples including Cuba, Venezuela, China and the Soviet Union show that socialists or communists can successfully capture state power for long periods and do a lot of good (as well as some bad) with it.

Some tendencies argue for approaches outside of the state because of its size. They might argue that a political body so large makes meaningful democracy difficult, produces an unwieldly and dispassionate bureaucracy, or still concentrates power in the hands of an only superficially different elite. However, the scale of the state is precisely why it is so crucial to climate justice.

These are all challenges we can work to resolve. Our task is to transform the entire economy, not just local communities. We need to transition the entire energy system from fossil fuels to renewables. We need a farming system which reliably feeds every person, not just areas with fertile land. We need a public transport system which connects every part of the country and beyond, integrating local, national and international travel. To meet the basic needs of every person, we need the universal provision of services. Scaling programs for climate justice to a national level doesn't preclude local initiatives. In many cases, national programs may be inspired by local initiatives, and the state can guarantee bottom-up democracy and resources for new co-operatives and local projects better than the market. Small may be beautiful, but in response to historic injustice and generational crises, it's not nearly enough.

The state is able to provide one thing that a private sector and market, local co-operatives and diffuse movements cannot: economic planning. Needless to say, transforming the economy is no easy task. It will require a centrally managed plan of targets, budgets, resource allocation, and key initiatives spanning years. In the US, the Green New Deal has been proposed as a ten-year plan for exactly this reason. The government will need to coordinate across the economy so that every sector can move in the same direction towards the common goals of decarbonisation and economic justice. The greater the state's capacity (that is, the resources at its disposal in the form of money, people, infrastructure, etc), the more effectively it can execute such a plan. State planning is also the best way to ensure that the upheaval is democratically directed. In the most obvious sense, governments are elected by publics. They receive a democratic mandate that way. But governments coordinating a Green New Deal would be able to coordinate a range of democratic forums to contribute to the planning process, including through citizens' assemblies, workers' councils, trade union representations and genuine public consultations. The state has the structures and the reach to guarantee formal mechanisms of wide-reaching democratic input, whereas the private sector serves only profit, and co-operatives and campaigns serve their members.

The private sector has dominated the economy in the years we've known about climate change, and over that time has not even tried to do anything to resolve the crisis. As the crisis becomes more urgent by the day, and even as capital begins to recognise (at least narratively) the usefulness of a liveable planet, there is no chance the private sector can be left to act with anything close to the urgency we need. On the other end of the spectrum, the co-operative sector is a marginal part of the economy which grows at a snail's pace. Other movements for social change are urgent in their response to grotesque and often fatal injustices, though many have been able to lean on the longer arc of history to make precious gains over many decades or generations. The imperative for climate justice is unique in that it comes with a timer. The longer it takes, the more difficult it gets. We're already in too deep, and each day we don't begin a radical transformation is one more towards irreversible planetary catastrophe. The state can drive the economic mobilisation of a Green New Deal forward by setting the pace with the right level of investment. Only the state has the fiscal capabilities to generate the funds for the levels of investment needed to transition the economy. Without the guarantee of profitability, private capital won't raise the funds needed itself. Local initiatives come nowhere close, often relying on community funding or member investment. The state isn't constrained by the need for short-term profits or restless shareholders. It has the power to tax and spend, produce some money itself, and run a budget deficit. States can invest knowing that they will receive returns in the long-term – not least in the form of a healthy population and stable climate.

Perhaps the biggest barrier to climate justice is the resistance of capital. It's no surprise that fossil fuel companies, aviation and banks spend so much time and money organising against legislation which would actually tackle the climate crisis. It's because they're the ones who will necessarily lose out. Climate justice requires marginalising the private sector and stripping the profit motive out of the economy. Unlike the state, grassroots or NGO movements and local co-operatives do not have the legal capabilities or resources needed to supersede capital in the economy. The state has the power to regulate, penalise or dismantle corporations. It has the power to nationalise companies or whole sectors of the economy or create new companies able to compete with private ones. That the state is the only political body with the power to confront capital is argument enough on its own for the need to use state power for climate justice.

Here I've painted an optimistic picture of the potential for states to contribute to building climate justice, compared to the alternatives. I don't do this because I naively believe states are unproblematic political formations or that state-led mobilisation of the economy to these ends are especially likely in our current context. This has demonstrably not been the case. I make the above arguments because those of us interested in climate justice need to be clear with ourselves that relying only on the private sector, co-operatives or localism, or the pressure of movements will be insufficient for the task. If we could direct state capacity to work for these ends, though, we at least give ourselves a chance.

## Who's in charge?

There are enough examples of state capacity being used to reproduce injustice, or simply of it being used poorly with good intentions, to justify a discerning approach to how the state is used in service of climate justice. I have made the case for the necessity of public ownership already, but we should certainly learn lessons from the past. Most claims by free marketeers that nationalised industries are fundamentally less efficient are baseless ideological claims. However, that such a logic was able to take hold as a foundational idea of neoliberalism suggests at least a kernel of truth. In the UK, some nationalised industries prior to their disastrous privatisation by Thatcher and Major had been starved of investment. There is little point in the state maintaining ownership of key sectors of the economy if they're not run to a higher standard than the private sector (which itself has a propensity to starve investment – look no further than Britain's railways or rental housing). I have already discussed the number of state-run fossil fuel companies which provide significant revenue to states, often funding welfare states and other public spending. That these national extractive ventures are necessary sometimes indicts the constraints of global fossil-fuelled capitalism more than individual governments. There are many states, however, that are actively reducing their dependency on these state enterprises, yet continue their operations. In many cases, there are opportunities not taken to lead a transition in the global energy sector by dismantling polluting industries at home. In the time that austerity and resurgent neoliberalism defined European politics, China has sought to out-develop the West by capitalist logics. But rather than drink the neoliberal Kool-Aid, China has undertaken huge investment domestically and internationally, expanding and exercising state capacity. One example of China's superior state capacity in action is its response to the COVID-19 pandemic. While the UK faced a third national lockdown, defining images out of China included people in Wuhan packing into clubs for New Year's Eve parties.[4] However, China's choice to try and beat the West at its own game has meant growing its geopolitical dominance with investment in high-carbon development projects across Asia, Africa

---

4   Danya Bazaraa, 'Wuhan locals party in packed nightclubs as much of world faces strict Covid lockdowns', *Mirror*, https://mirror.co.uk/news/world-news/wuhan-locals-party-packed-nightclubs-23245202 (accessed 30 January 2021).

and into Europe. That so much of this investment is directed to new coal mines and power stations emphasises that the climate crisis is at best only a secondary concern to the Chinese state seeking to succeed within the constraints of global capitalism.

So it matters who is in charge of the state both to ensure that we have sufficient state capacity and that it is used for the right ends. Those of us committed to climate justice should get serious about strategies for capturing state power. We need governments prepared to use state capacity to coordinate a huge economic mobilisation of mass investment, democratisation and transformation. We need governments able to join the dots of climate justice too by diverting the state capacity channelled into policing, prisons, military, detention and deportation – presently used for racist violence, mass incarceration, and war – towards meeting basic needs, real justice and decarbonisation.

## The strike and the ballot box

Practically, how do we capture state power to use it for these ends? On the timescale we have, there are broadly two options. We can win state power through democratic elections or seize it through a revolution. We're not in a position for half measures here. Whichever of these two strategies we choose, we should really go all in on it. A revolutionary strategy can be appealing if you have little or no faith in the institutions of capitalist democracy. Leftist electoral movements of the 2010s have largely been quashed. Even in so-called liberal democracies the system is rigged in favour of the ruling class. You might say our only realistic choice is to overthrow the government and take state power by force. You can certainly see the logic to this line of thinking, but pulling off a revolution is no easy task. If this is your chosen strategy, you should be prepared to use violent force because the capitalist state certainly won't hesitate to use it against you. You'll probably need some method of arming your supporters. You should be confident in being able to win majority support among the military and police (ideally both). You should be prepared to occupy key infrastructure like airports, broadcasting and energy production. Thinking about this strategy in the context of capitalist democracies – among the most heavily militarised surveillance states history has ever seen – all of a sudden, the prospect of revolution feels more distant.

In capitalist democracies, our only strategic option is to build a mass democratic movement for climate justice to win state power at the ballot box with the support of radical social movements and militant trade unions. Environmental movements' tendency towards marginality has meant the work of building such a mass democratic movement has largely been neglected over recent decades. So there's work to do, and fast. Seeking to win state power through the electoral route will mean organising through political parties. The question for the climate justice movement is: which parties? Things to consider include which progressive parties have a chance of winning power within the domestic electoral system; in which parties can the climate justice movement exert most influence over the policy platform; can the party have mass appeal; and is there scope to build a viable new party to meet the movement's electoral needs? In the UK's context, the Labour Party is the only party with a chance of forming a government under the first-past-the-post voting system. It has an imperfect internal democracy, but there is scope to elect a radical leadership (see Jeremy Corbyn's elections) and shape the party's platform. In the US, the choice of parties is even more constrained. So far, climate justice movements have organised through the Democratic Party taking advantage of its primary system to elect progressive leaders to

Congress and other positions, with two runs for the presidency by Bernie Sanders. In European contexts where proportional representation is less restricting, there is more scope to form new parties or at least not hitch the movement to declining centre-left parties.

Activists have a track record of grassroots organising to shift the climate platforms of both established and newer parties. At the 2017 general election, Jeremy Corbyn's Labour Party promised to ban fracking, insulate four million homes to end fuel poverty, and bring energy into public ownership. By the 2019 general election, Labour for a Green New Deal's campaigning successfully shifted Labour so that its plans for a green industrial revolution were front and centre of the manifesto, including the target to decarbonise almost entirely by 2030, a greater level of investment ambition, insulation for every home in the country, and stronger public ownership commitments covering the national grid and big six energy companies, and broadband internet.[5]

Bernie Sanders' presidential campaigns progressed similarly. In 2016, Sanders ran against Hillary Clinton to put forward a radical alternative to her faux-progressive pro-corporate, war-mongering commitment to upholding business as usual. Sanders pledged to ban fracking, to tax carbon and he supported Indigenous resistance to the Dakota Access Pipeline at Standing Rock. By 2020, Sanders embraced the Green New Deal framework, which included a pledge of $16.3 trillion of public investment and detailed plans for a just transition for workers in affected industries. Sanders went after the fossil fuel industry, calling for executives to be held to account for their role in the

---

5   'It's Time for Real Change: Labour Party Manifesto 2019', p. 12, https://labour.org.uk/wp-content/uploads/2019/11/Real-Change-Labour-Manifesto-2019.pdf (last accessed 17 April 2020).

climate crisis, emphasising the need to transition from fossil fuels to renewables, and making big polluters pay. This came after grassroots campaigning by the Sunrise Movement, and the popular energy that Justice Democrats created around the Green New Deal.

In Spain, Podemos' initial platform paid limited attention to climate justice, with the recent financial crisis and austerity measures still dominating European politics. By the April and November 2019 general elections, Podemos' climate platform included policies to transition to 100% renewable energy by 2040; end nuclear and coal generation by 2025; tax industrial polluters; and insulate half a million homes.[6] Txema Guijarro García, a Podemos politician and economist, told *Jacobin* magazine that Podemos' 2019 platform was a response to pensioners' strikes, feminist mobilisations and the climate protests that had taken place over the last year. They planned to create a public energy company and a state investment bank to secure the energy transition, using the Green New Deal framing to articulate their focus on tackling climate change and Spain's structural unemployment through investment. While both Corbyn and Sanders ultimately failed to win their elections, Podemos' share of the vote declined but formed a coalition government with the centre-left PSOE. Highlighting these examples isn't about harking back to failed electoral movements that could have been, but demonstrates that political parties aren't static organisations. With an active, well-organised and radical membership combined with outside pressure from social movements, we can prepare progressive parties to take state power and implement a Green New Deal.

---

6   Richard Weyndling, 'Spain's new government promises climate will be a priority', *ENDS Europe*, https://endseurope.com/article/1665740/spains-new-government-promises-climate-will-priority (last accessed 20 April 2020).

# THE S-WORD (STATE POWER)

It's only by acknowledging the limitations of capitalist democracies, the flaws of centre-left parties and failures of the recent past that we stand a chance of overcoming them. We know that the billionaire-owned media promotes the propaganda of the ruling class and electoral systems are designed to keep them in power. We know that centre-left parties are generally controlled by careerists and bureaucrats more interested in maintaining their control of the party than winning state power. We know that there is no electoral shortcut to socialism in Western countries with powerful internal opposition. We know that if we do manage to leap over all these hurdles to win at the ballot box, the military, civil service and capital will throw everything they have at disrupting any radical government. These are the reasons we can't hope to win state power only by organising through political parties, essential as the electoral route is.

We need radical social movements and a militant labour movement prepared to fight to win state power and then defend it. From opposition, these extra-parliamentary movements can use their power to exert influence on the levers of state power while they are still held by the ruling class. Through mass mobilisation, direct action and industrial action they can begin to win concessions for workers' pay and conditions, transitions out of polluting industries and investment in green alternatives. These movements also have a crucial role to play in creating the conditions for a radical party of climate justice to win at the ballot box. Through this action, radical movements will build popular support for a transformative Green New Deal and expose the ruling class as incapable of dealing with the intersecting climate, economic and public health crises we face. Trade unions could fund alternative media projects to counteract the dominance of the billionaire press and shift the public debate to build support for a socialist vision of climate justice. When our movement does win state power, those same radical social movements and militant trade unions will need to mobilise in defence of our radical agenda, pushing back against the resistance of capital. It will also need to push any radical government to go further with bold action, acting as a counterweight against the conservative tendencies of government. So even if organising in elections and political parties isn't your cup of tea, there's still so much to do in the struggle to win state power and use it for climate justice.

# Chapter 8

# Don't let crises go to waste

It's an interesting experience to write a book about crises, injustice and movements at the same that another crisis emerges and develops – particularly when it intersects so clearly with those others while turning our usual methods of activism upside down. As if the climate crisis and deep injustices of capitalism weren't enough to deal with emotionally, intellectually or practically, another crisis emerged when I began writing the book. Enter COVID-19.

The COVID-19 pandemic and responses to it shut down economies, left many unemployed, pushed healthcare systems to the limit, put social lives on pause, and took millions of lives. From the early stages of the pandemic, it was clear to many of us that it was impossible to separate COVID-19 from climate change – both the root causes and the solutions they require. NGOs, trade unions and political parties called for a 'green recovery' or to 'build back better'. The idea was for governments' economic recovery plans to integrate key measures from the Green New Deal to tackle deeper climate and economic injustices in tandem. The alternative would be a 'recovery' which locks us into another decade of business as usual. This opportunity was largely wasted by governments. Some countries with greater state capacity and pandemic preparation got the virus under control relatively quickly. Other governments oversaw humanitarian disasters, sacrificing many thousands of lives in efforts to preserve profits over human lives. Regardless of how quickly governments dragged their economy back to relative normality, it was not enough. 'Normal' in this context only represents a return to the conditions which produced these crises in the first place.

## Pandemic and climate

It's likely that COVID-19 jumped from a bat to a pangolin and then to people at a market in Wuhan, China. This is an example of 'zoonotic spillover' where viruses move from animal populations into human populations.[1] Zoonotic spillover is made more likely by a combination of factors which are striking in their similarity to some of the root causes of climate change. Industrial agriculture and deforestation are well known causes of emissions and biodiversity loss. They can also be blamed for bringing wildlife, previously spreading pathogens only between each other, into close contact with human populations. Accelerated global transportation is another great source of emissions and pollution. This played a key role in making the COVID-19 outbreak into a global pandemic, as carriers of the virus brought it from continent to continent.

Austerity, privatisation, fortified borders and precarious work are making us more vulnerable to the effects of climate change. These economic pressures also cause us to experience the pandemic as a disaster, with states stripped of the capacity to respond effectively.[2] We have already experienced three distinct

---

1   Andreas Malm, *Corona, Climate & Chronic Emergency* (London & New York: Verso Books: 2020), p. 31.

2   Malm, *Corona*, p. 102.

coronaviruses in the twenty-first century: SARS, MERS and COVID-19. At present there is no sign that governments are willing to take the measures that would cut these out at the root. If we stick with the status quo of industrial agriculture, mass deforestation, accelerated global transportation and an erosion of state capacity, we will face even more disasters on an even greater scale over the coming decades.

It has been those industries, stalwarts of global capitalism, most responsible for emissions, that were among the most disrupted by the pandemic's economic shutdown. In April 2020, the price of a barrel of oil briefly dropped below zero, priced at minus $37.63 at its lowest point. This was an example of disruptive real-world events getting in the way of real-world assets being traded on financial markets by speculative investors. Usually, traders don't have to deal with the barrels of oil they're passing between each other via computers. When demand collapsed amid economic shutdowns, traders realised they may actually have to receive and store the asset – which they obviously didn't have the facilities to do – so ended up paying to get rid of it. Although the price soon crept above zero again, the structural issues of oil markets weren't resolved. The same thing could easily happen again with serious implications for markets.

The economic shutdowns have afflicted the aviation industry to an even greater degree. As commercial flights were grounded, several airlines went bust and others had to make severe cutbacks. Norwegian Air cancelled 85 per cent of flights and laid-off 90 per cent of staff. Lufthansa Group laid off 22,000 staff just weeks after a €9 billion bailout from the German government. British Airways cut over 8,000 jobs.[3] In the auto-

3    Jim Armitage, 'British Airways job cuts top 8,200 as parent company IAG raises billions of euros to shore up finances', *Evening Standard*, https://standard.co.uk/business/british-airways-jobs-redundancies-travel-quarantine-covid-a4544146.html (last accessed 1 February 2021).

motive industry, manufacturing ground to a halt as demand slumped and social distancing measures forced factories to shut. Halfway through 2020 the UK industry predicated that one in six jobs in the industry would be lost.[4] The question around these industries is whether we will see a managed, just transition away from them, or whether they will collapse under the pressure of crises they have themselves produced (causing only more disaster).

That the pandemic afflicted the most polluting industries in the context of generally depressed economic demand meant a temporary reduction in emissions. 2020 saw a 7% reduction in emissions. This was the first drop since 2009 (then caused by the global financial crisis) and the largest absolute drop ever recorded.[5] The fact that only major crises endogenous to capitalism have had any notable impact on emissions is an indictment of efforts (or lack thereof) to decarbonise. Needless to say, the economic collapse, unemployment crisis, long-term health impacts and mass deaths aren't a price worth paying for a temporary blip in otherwise relentlessly rising emissions. Nor do they represent a model for full decarbonisation. The effect of the pandemic and financial crisis on emissions only underlines the insufficiency of decarbonisation strategies based on simply 'reducing emissions' or limiting economic activity. We need to eliminate emissions by changing the energy base of the economy (from fossil fuels to renewables) and transforming how the economy operates (from private ownership for profit to public ownership for justice).

---

4   Gwyn Topham and agencies, 'UK car industry "could lose one in six jobs due to Covid-19 crisis"', *Guardian*, https://theguardian.com/business/2020/jun/23/uk-car-industry-jobs-due-covid-19-smmt-redundancies (last accessed 1 February 2021).

5   'Global Carbon Project: Coronavirus causes "record fall" in fossil-fuel emissions in 2020', *Carbon Brief*, https://carbonbrief.org/global-carbon-project-coronavirus-causes-record-fall-in-fossil-fuel-emissions-in-2020 (last accessed 1 February 2021).

The pandemic's temporary emissions reduction is a good headline, but more significant is the effect of the pandemic on climate activism. The new wave of grassroots climate organising (discussed in Chapter 4) was growing in momentum until the pandemic hit. Suddenly it was flattened as national lockdowns limited the available forms of mobilisation. Events moved online. Youth strikes stopped. XR paused all mass action from March 2020, prior to a planned period of rebellion in May and June. At that time co-founder Gail Bradbrook made public appeals for funding, claiming the organisation needed £250,000 to survive due to the costs of cancelling actions at such short notice.[6] By the end of 2020 and at the beginning of 2021, XR began to organise in-person mobilisations again. Climate organisations were rendered relatively toothless at precisely the moment they needed to be piling on the pressure for government to put climate justice at the heart of its recovery plans. The lack of preparedness for this unprecedented situation isn't really surprising. However, we can learn the lessons of this pandemic and get serious about making the most of the opportunities that present themselves. Future disasters are inevitable. We need to be ready to respond to defend ourselves both in the immediate and in the long term. We need to protect each other in moments of extreme vulnerability, while using them to bring radical demands and force bold action – regardless of the challenges of the situation.

## Twenty-first century breakdowns

Capital has already shown its readiness to take advantage of these crises. Corporations have attempted to fire and rehire

---

6   Catherine Early, 'XR needs £250,000 to survive', *Ecologist*, https://theecologist.org/2020/mar/23/xr-needs-ps250000-survive (last accessed 5 February 2021).

workers on worse pay and conditions. In the UK, workers at British Gas and Heathrow Airport went on strike to oppose plans.[7] British Airways had planned a round of firing and rehiring before pressure from unions forced them to back down.[8] Fossil fuel companies have used the pandemic to lobby for their own interests to be baked into 'green recovery' plans across Europe. In Portugal, António Costa Silva, the chief executive of fossil fuel company Partex, was appointed to write the country's economic recovery plan having previously criticised the government for blocking new oil and gas exploration in the Algarve.[9] At the EU, lobbyists met with high-level officials 25 times on behalf of fossil fuel companies between 23 March and 25 May 2020. They were promoting deregulation and support for fossil fuels in the EU's COVID-19 recovery package.[10] Truly, the hustle never stops.

Grassroots movements need a strong plan for responding to the disasters we will see over the coming decades because capital certainly has one. We need to make sure our movements aren't immobilised but are ready to adapt to new contexts. While climate mobilisations dwindled during the pandemic, there was a surge of community-based solidarity to meet the needs of those most vulnerable during the pandemic. Mutual aid networks were set up in towns and cities across the UK.

---

7 'British Gas staff start five-day strike in "fire and rehire" row', *BBC News*, https://bbc.co.uk/news/business-55562904 (last accessed 6 February 2021).

8 'BA boss says there is no need to fire and rehire staff', *BBC News*, https://bbc.co.uk/news/business-54176543 (last accessed 6 February 2021).

9 Chris Saltmarsh, 'How Fossil Fuel Companies Have Exploited the Covid Pandemic', *Tribune*, https://tribunemag.co.uk/2021/01/how-fossil-fuel-companies-have-exploited-the-covid-pandemic (last accessed 6 February 2021).

10 Francesca Newton, '10 Ways Corporations Have Exploited Covid-19', *Tribune*, https://tribunemag.co.uk/2020/12/10-ways-corporations-have-exploited-covid-19 (last accessed 6 February 2021).

With some people forced to isolate in their homes and others running low on money due to layoffs, the networks worked to distribute food and other necessities. These networks plugged the gaping holes left in the state by decades of neoliberalism. While the government failed to offer proper support, neighbours came together to provide for each other. Local mutual aid networks are not a sufficient alternative to state support during these shocks; meeting basic needs should not really be dependent on the organisation, resources and capacities of local volunteers. As climate change brings more extreme weather events, flooding, drought, and wildfires we will see even more disruptions requiring emergency response. We need to demand programs of just adaptation to climate change and pandemic response plans, at the core of expanding state capacity so that governments can reliably provide emergency rescue services, quarantine facilities, food, shelter, healthcare and income. We should make these demands now, but we cannot rely on the state. Movements need to be ready for disasters in which the state is not prepared. Over the coming decades, mutual aid networks will be as crucial a part of radical climate movements as mass marches and direct action. We need to be ready to keep each other safe while negligent governments are happy to sacrifice us. We need to get organised now because when the disaster hits there'll be no time to lose.

In Manchester, Green New Deal activists responded to the pandemic by setting up a cooperative project called Retrofit Get-in. It gave laid-off theatre workers jobs using their skills to retro-fit homes to make them carbon neutral. Theatre workers were hit hard by the pandemic. One of the project's founders, Andrew, 'was 12 hours away from the get-in of a show he had spent two months working on when the UK went into lockdown; within two weeks the rest of his work had been cancelled or postponed until 2021.'[11] The unemployment crisis will not be solved by local cooperative projects on their own. What we need is massive investment in a program of green jobs with guarantees of work (with equivalent or improved pay and conditions) for anyone working in polluting industries or those disrupted by pandemics and other disasters. However, projects like Retrofit Get-in are crucial first steps towards that goal. They provide a platform from which to make these larger scale demands. These local initiatives make the key elements of a Green New Deal look both possible and necessary. They expose the failures of government and our business-as-usual economy when they perform what should be the role of the state. Drawing inspiration from these Manchester activists, we should do more of the work of prefiguring what a Green New Deal could look like for workers and communities. Local projects are another method, other than marches and direct actions, to create a sense of possibility and eventually inevitability around the new society we need to see.

Although the pandemic quashed mass mobilisation around climate change, it saw a resurgence of the Black Lives Matter movement. Millions of people around the world took to the streets in protest against police violence and racism after the killing of George Floyd by Minneapolis police.[12] This was at the height of the initial lockdowns in May and June 2020. The uprisings were clearly driven by emotion, whether it was rage, grief or hope. That character caused me to reflect on the

---

11 'How Laid-Off Theatre Workers Are Tackling the Climate Crisis', *Tribune*, https://tribunemag.co.uk/2020/10/how-laid-off-theatre-workers-are-tackling-the-climate-crisis/ (last accessed 6 February 2021).

12 'Black Lives Matter May Be the Largest Movement in U.S. History', *New York Times*, https://nytimes.com/interactive/2020/07/03/us/george-floyd-protests-crowd-size.html (last accessed 7 February 2021).

routine of climate protest. Not for a long time had I felt the same emotion on a climate demo. When was the last time I stood among fellow activists and felt the palpable rage at the overwhelming injustices of the climate crisis? I think this emotion was present at early youth strikes and XR rebellions, but that quickly fades away. As climate activists, we've slipped into a ritualistic approach to protest and mobilisations. We hold them so regularly that they are no longer a powerful expression of shared emotion or desire. That these protests became so routine meant that they stopped during the lockdown just like other everyday activities like going to the pub, the cinema or a friend's home.

It might be worth climate activists keeping mass marches, protests, and social disruption in our back pockets a little more. They shouldn't be our default form of action, but a powerful weapon to deploy at key moments. In between, we could prioritise the longer work of building power by organising communities and workplaces, building local and international solidarities, and working through political parties and trade unions. Then, when a pandemic shuts down the economy, when the next wildfires hit, when extreme weather rips apart entire cities, when flooding destroys homes, or when drought disrupts the food system, we can explode onto the streets better organised with powerful demands and greater moral authority. We can force the hand of the ruling class as its scrambles to deal with yet another crisis of its own making. We can move a step closer to taking power ourselves.

# Conclusion:
# Don't mourn, organise!

The environmental NGO industry will spin anything as a victory: fossil fuel companies publishing net-zero ambitions which don't include their actual extraction; inadequate international accords like the Paris Agreement; or banks excluding finance for projects it didn't support in the first place. NGOs seem to rely on a constant churn of victories to create a sense of momentum, even when it's not really there. Part of this is to remain positive in the face of climate catastrophe. It's not easy to devote your entire professional life to a crisis of this magnitude. Part of it, though, is also to justify the industry's existence. Convincing each other that the work that we do is making a difference. Too often, it isn't. This relentless positivity emanating from the NGO industry has contributed to a collective delusion that we're on the right track to stop climate change. Unfortunately, this couldn't be further from the truth. Emissions continue to rise unaffected by NGO campaigns, international climate conferences, direct-action blockades and marches in the street. Flooding, drought and extreme weather continue to increase in frequency and severity, spreading to new geographies at the same time. Many people have already lost their lives to climate change or been displaced from their homes. Many, many more people will experience the same fate in the coming decades. I don't want to indulge in false optimism about our chances of achieving justice because we need to be realistic about the challenge facing us and what it will take to overcome it.

## CONCLUSION: DON'T MOURN, ORGANISE!

I hope that you finish reading this book with a clear sense of the scale and nature of the climate crisis: that it means the most horrifying devastation to the lives and livelihoods of those people in every part of the world who have done the least to contribute to its cause. I hope you go away with a clear sense of who has caused this crisis: the ruling class, the super-wealthy, and the capitalist system that works in their favour. I want you to feel that no matter how painful and overwhelming it is to know this, we do still have a chance in our fight for climate justice. It will take everything we've got. The antidote to confronting these difficult truths should not be depression or hopelessness. At this point, this is just the hand we've been dealt. We don't have to be optimistic (on most days I'm not), but we can be hopeful. We can draw inspiration from the successes of our comrades struggling for justice throughout history and from the new political generation just beginning its journey. We carry the torch of that struggle not because climate justice is inevitable, but because it is possible. And this possibility is not just business as usual with a stable climate. Climate justice is shorthand for a new society and economy which guarantees prosperity for all people, not just the wealthy few. As much as it is an existential threat, the climate crisis is a historic opportunity.

So what can we do to take this opportunity? It's time to get organised.

Are you best placed to take part in direct action confronting capital? Can you organise mass mobilisations of young people in the streets? Can you target politicians to support a Green New Deal? Will you demand transformations to the financial system? You could organise community mutual aid networks or start initiatives that put the principles of a Green New Deal into practice locally. Whatever you choose your contribution to be, remember that your key aim should be to exercise collective power to force the levers of state power to act for climate justice. Seek to create the conditions for socialists to take state power by popularising a radical vision for climate justice. Be prepared to both defend and push further those in power committed to climate justice.

Join a union, but don't just leave it at that. Organise your workplace and fight to win concessions for you and your colleagues. Build consciousness around climate injustice among your fellow members. Agitate for your union to prioritise the fight for a just transition. Agitate for the union to support militant industrial action on both workplace issues and national politics. A Green New Deal has to be led by a militant labour movement. It's up to workers everywhere to make that a reality.

Join a political party, but not as a moral virtue or expression of identity. Choose a party with a chance of winning power and the political space to adopt a socialist Green New Deal platform. Treat parties as a terrain of struggle and a vehicle for achieving climate justice through the state. Organise with fellow members for your party to adopt the boldest climate justice platform possible, threading it through every other issue. Build a powerful movement to win power, and make sure you're ready for when that day comes.

Now is not the time for moderation in politics or strategy. The climate crisis commands that we build a radical and militant movement for a huge economic transformation. It won't be easy. The clock is ticking. We had better get on with it.

# Resources

## Groups

Coal Action Network: coalaction.org.uk/
Ende Gelände: ende-gelaende.org/en
Friends of the Earth Scotland: foe.scot
Fuel Poverty Action: fuelpovertyaction.org.uk
Gastivists Network: gastivists.wordpress.com/
GMB for a GND: twitter.com/gmb4gnd
Health for GND: twitter.com/Health4GND
Indigenous Environment Network: ienearth.org
Labour for a Green New Deal: labourgnd.uk
London Mining Network: londonminingnetwork.org
Momentum: peoplesmomentum.com
People & Planet: peopleandplanet.org
Reclaim the Power: reclaimthepower.org.uk
Sunrise Movement: sunrisemovement.org
Tipping Point: tippingpoint.org.au
UK Youth Climate Coalition: ukycc.com

## Initiatives

Banking on Climate Change: ran.org/bankingonclimatechange
2020/
Find your union, TUC: findyourunion.tuc.org.uk
Global Campaign to Demand Climate Justice: demandclimate
justice.org/

La Via Campesina: viacampesina.org/en/

The Leap Manifesto: leapmanifesto.org/en/the-leap-manifesto/

Manchester Green New Deal Podcast: podcasts.apple.com/gb/
podcast/manchester-green-new-deal-podcast/id1501980017

Offshore: Oil and gas workers' views on industry conditions
and the energy transition

One Million Climate Jobs: Tackling the Environmental and
Economic Crisis: cacctu.org.uk/climatejobs

Peace & Justice Project: thecorbynproject.com

Perspectives on a Global Green New Deal: global-gnd.com

platformlondon.org/p-publications/offshore-oil-and-gas-
workers-views/

Retrofit Get-in: retrofitgetinproject.com

Stop Adani: stopadani.com

The World At 1°C: worldat1c.org

Thanks to our Patreon Subscribers:

*Lia Lilith de Oliveira*
*Andrew Perry*

Who have shown generosity and
comradeship in support of our publishing.

Check out the other perks you get by subscribing
to our Patreon – visit patreon.com/plutopress.

Subscriptions start from £3 a month.

**The Pluto Press Newsletter**

Hello friend of Pluto!

Want to stay on top of the best radical books
we publish?

Then sign up to be the first to hear about our
new books, as well as special events,
podcasts and videos.

You'll also get 50% off your first order with us
when you sign up.

Come and join us!

Go to bit.ly/PlutoNewsletter